Robert Sharrock

The History of the Propagation and Improvement of Vegetables

Robert Sharrock

The History of the Propagation and Improvement of Vegetables

ISBN/EAN: 9783337375300

Printed in Europe, USA, Canada, Australia, Japan

Cover: Foto ©berggeist007 / pixelio.de

More available books at **www.hansebooks.com**

THE
HISTORY
OF THE
Propagation & Improvement
OF
VEGETABLES
By the concurrence of
Art and *Nature*:

Shewing the several ways for the Propagation of Plants usually cultivated in *England*, as they are increased by Seed, Off-sets, Suckers. Truncheons, Cuttings, Slips, Laying, Circumposition, the several ways of Graftings and Inoculations; as likewise the methods for Improvement and best Culture of Field, Orchard, and Garden Plants, the means used for remedy of Annoyances incident to them; with the effect of Nature, and her manner of working upon the several Endeavors and Operations of the ARTIST.

Written according to OBSERVATIONS made from Experience and Practice:

By *Robert Sharrock*, Fellow of *New Colledge*.

Oxford: Printed by *A. Lichfield*, Printer to the University, for *Tho: Robinson.* 1660.

TO THE
HONORABLE
ROBERT BOYLE *Esq*;

The most worthy pattern of true Honor,

AND

Learned Promoter of true Science.

SIR,

IT is a saying in the Civil Law, *That a thing which is any Mans own, cannot be made more his by any new Act or Deed:* The consequence of which, is, that the Dedication of this Piece to you will be meerly nugatory, since by all right it is already yours. For it is not long since I imagined no more being either Author, or Compiler of any matter on this subject, then of doing any other thing which I have neither fancy nor fitness to. But you were pleased to judge me able, and (which obliged me to this task) to propose it unto me as your de-

The Epistle Dedicatory.

sire that I should make an essay of that ability, in writing somewhat even on this subject, that might be of Philosophical and common use. To have questioned your judgement herein, might have stained me with too much arrogance, and to have been carelesse of your pleasure, with unworthiness and want of good Manners. Remembring therefore those respects I owe to Honor, Learning, and such persons as study its advancement and promotion, I could not deny this poor endeavor, the product of which arising originally from your own act, I thought fit should be delivered over to your pleasure, since to you, as its primary cause (which is its prime commendation) it ought to belong.

And Sir, If it may not be troublesome unto you to receive some brief account of this action, and the Fortunes which happened to me in pursuance of your satisfaction therein, you will give me leave to acquaint you, that it having been your Honors express desire, that this Piece might extend as far, and be as comprehensive and full, as my present Experience, Knowledge, and Recollection of the matter of Vegetable Propagation should permit: I gave my self the trouble to run over with my eye, all Books I could procure of these subjects, not intending to trust any, but thereby to be put in mind

of

The Epistle Dedicatory.

of the particulars, concerning which, I had no reason to have a Register ready in my head. Here first my fortune was to finde a multitude of monstrous untruths, and prodigies of lies, in both Latine and English old and new Writers, worse in their kinde then the stories in Sir *Iohn Mandevel's* Travels, or in the History of Fryer *Bacon* and his Man *Miles*; or else what may be more ridiculously removed not onely from truth, but from any semblance thereof. And which moved me most at this very season, when we esteemed the World to be now awaked, I found in the Shops Authors newly set forth (I hope against their own wills) who seriously professed to have *made a select choice of Experiments* of this nature, *and to report nothing, but what from observation and experience they have certainly found true,* yet deserving not to have the credit of *Wecker* and *Porta*. Professions in such Papers, which seem to me at no time proper, but when the persons credits, together with their Books, are joyntly to be set to sale. You easily believe that I am not free to follow these Examples, for then, first, I must abuse your Noble Name, by inscribing it to a most unworthy Discourse, and then (which is too common a fault) traduce as many Readers, as ignorance and simpleness of nature hath made credulous.

But as to those Authors, in whose relations I found any thing of truth, I have done them this right, That where ever I could relate an Operation or Experiment in their words, with truth and fitness, I spared to coyn new (desiring to supplant no Author in his credit, nor to purloyn his reputation) though I had learned the truth of the same thing from the testimony of my eyes: Having indeed some quarrel at the fashion of ordinary Writers, who study in nothing to benefit Learning, but by giving new words to old matter.

I have left out none of the Heads proposed in the Catalogue, which I presented you with, a year since, except the last, which you desired might shew the methods and ways of keeping useful Vegetables without putrefaction, and the preparing them with their several parts and products for humane use. This at present I thought necessary to forbear, for I found the matter too much for one Chapter, and my leisure too little to make a Book thereon: nor durst I esteem my Observations such, as might enable me to write an adequate Treatise on that subject, which reaches in compass the largest, and as I firmly believe (however the Animal and Mineral Kingdom abound with
great

The Epistle Dedicatory.

great and potent Medicines) not the worst part of the *Pharmacopœa*, and many particulars beyond; but rather think fit to employ my self some more years in the Experience and Practice of Preparations, and take the pains of collecting and trying such intelligible and probable processes as shall come to my hand, either reported heretofore, or used now, especially in our Nation, for fitting matters to Alimental, Medical, and Mechanical use, before I shall imagine to have the least hand in that History, which may as well be learn'd by such as are concern'd to know it, from Modern Dispensatories, and other novel Writers. But the perfection of that History, with correction of processes capable of amendment, is, in my estimation, a design and work worthy of the Care, Patronage, and Governance, and fit to be carryed on by the interest, if too tedious, for the Pen and Pains of your Honor.

As to the form and composure of matter under those Heads, I must make it a particular business to beg your pardon; for I finde it even in my own judgement exceeding rude, and it could be no otherwise, when the Revise of the Press, was, for a great part, the first review made of my own Writing; and indeed, the whole piece in every

every part seems destitute of beauty, and without any thing of great worth, value, or nobleness. For I finde, that the operations themselves, and other matters that do belong to the subject in hand, and so capable to come under this History, are for the most part common, and devoid of curiosity: Nor durst I embellish their plainness with Stories taken from our Learned and Profound Writers of Natural Magick, because I intended, as no very imperfect, so likewise a true Inventary of what the power of man, at this present time, on this subject, is, with the Co-operation of Nature, able to produce: For these reasons, and perchance because of another piece then under my hand, to which I had more propense affections. I was exercised in this writing, not without some reluctancy and untowardness of minde; and it surely had proved to me a piece of meer drudgery, had not the hope of giving you satisfaction, and making this a testimony of my obeysance and humble submission to your Judgement and desires inspirited me, and let a lightsomness into my thoughts. What I have written, I shall not commend, by any Prefaces, to any Reader, though I shall give him here some things new, and of my proper Observation: I know that many, by their own Interest

and

The Epistle Dedicatory.

and (that great power) Temporal Profit, will be tempted to give it the reading. Neither shall I, in imitation of some Modern Alchymists, for ostentation, bid them goe; and by the improvement (which I hope may be some to most Readers) be charitable to the poor: Hoping, that for Gods sake, they will rather (as they are bound by Obligations infinitely more high) be thereto moved; nor need I excuse my self to them for any deficiency in this Writing, you having ingaged your self to be the Proprieter thereof, and by your acceptance of this poor Piece, greatly obliging,

SIR,

Your Honors unfeignedly Devoted in all humble and affectionate observance.

R. SHARROCK.

To the Author on his two late publisht Pieces, *The Hypothesis of the Law of Nature*, and, *The History of Propagation*.

SIR,

OF late to th' privy Chambers of the minde
You led's, to which a glimmering ray had shin'd
From God th' Abyss of light; but much adoe
There had been made to stop that ray out too.
Here 'twas you drew a Curtain, and we saw
The sacred Tables of our Natures Law;
The frame of which was made of polisht glass,
Where each Soul, fair and foul, might see its face:
And there hung Justice Scale, ready to weigh
All actions, good and bad, just as they lay:
Justice her self we saw not, for 'twas sed,
That long agoe her Ladyship was fled.
But Duties way-marks, up and down there stood,
And the forgotten bounds of Ill and Good.
Much Furniture besides; all by th' abuse
Of new invented fashions, out of use.

Now Sir, you'r walkt abroad, you teach to Sow,
And Plant, and Graft, and shew how all things grow
By th' best improvements; how to harness Art
With Nature, and to make her draw her part:
How Nature varies all her Scenes, and makes
Things orderly and useful for our sakes.
You trace her steps, and make us plainly see't,
To be great Providence that guides her feet:
Thus when at home, and when abroad, you can
Contrive to honor God, and pleasure Man.

Will: Parker, Scholl: of *New Coll.*

A Gratulation unto the Author, upon his History of the Propagation of Vegetables.

WEe'l blame Antiquity no more, that she
   Has swallowed *Solomons* Phytology;
Those long-lost sacred Relicts you revive,
Limning the nature of each Vegetive.
Natures most hidden store, you open set,
As if y'were keeper to her Cabinet.
  Mid'st Plants and Trees you muse, thence we confess,
England again hath got her *Druides*:
Your Garden, a new Academy; can
Be made *Lycæum*, or turn'd *Vatican*.
  So the fam'd Epicure, long since did try
To make his Garden teach Phylosophy;
Where he, by *shuffling* Atomes, represents
All changes; a *Cator* of Elements
He then *laid out*, and (what was yet more high)
Boldly *discarded* Heavens Deity.
You slight that *play*, and shew there's no *sequence*,
No *suit of things*, without a Providence.
Each Herb's engraven'd with a Heavenly Frame,
Like th' *Hyacinth* enstamp'd with *Ajax*'s name:
As a mysterious Rabbin's wont to spell
The name of God, from a dark Syllable:
So you read him in's secrets works; Each clod
Speaks th' God of Nature, makes not Nature God.

  May these your Vegetives, thus ordered, prove
   A *Vocal Forrest*, or *Dodona's Grove*.
  To speak your worth, that so our non-plus'd cry
  May be assisted by *Dendrology*.

                Ed: Spencer Fell: of *N.C.*

The CONTENTS.

CHAP. I.
Of Propagation by Seed.

Pag.

Num. 1. *Of propagation of Vegetables in general, with a Preface to the Discourse.* 1

Num. 2. *A Catalogue of Plants that may be encreased by Seeds: with a question touching Maiden-hair, Harts-tongue, and Plants of like nature.* 4

N. 3. *The Seasons of sowing particular Plants, with proper Animadversions to this head.* 10

N. 4. *Examples of Sowing, with some particular directions for some choice Vegetables; with general Observations for the manner of sowing.* 16

Examp. 1. *From Mr.* Parkinson, *directing skilfully the ordering of* Tulips, *in their propagation by Seed.* ibid.

Examp. 2. *Of Anemones.* 18
Examp. 3. *Of Clover-grass.* 19
General Observations for the manner of Sowing. 23

Num. 5. *Of variety of kindes, different in colour, taste, smell, and other sensible qualities, proceeding from some seeds, and what Plants they are that bring seeds yielding such variety, whence the beauty of Flowers chiefly arises.* 25

Num. 6. *Some other relations touching transmutation, and the possibility of a change of one Species into another,*

CONTENTS.

ther, *examined in particulars of the Vegetable, Animal and Mineral Kingdoms.* 28

N. 7. *Of preservation for Seed, with advantagious directions therein.* 33

N. 8. *The manner of growing by Seed, Historically set down, with some Philosophical conceits thereabouts.* 35

Num. 9. *Of the cause of greenness in the Leaves of Vegetables.* 40

CHAP. II.
Of Propagation by Off-sets.

Num. 1. *A Catalogue of Plants which may be propagated by Off-sets and Suckers arising with Roots from the Stool and Root of the Mother Plant.* 43

N. 2. *The way of making Off-sets by Art.* 45

N. 3. *Rules for direction in taking off Suckers, or Off-sets.* 46

N. 4. *Examples of planting by Off-sets in Licorice, Hops and Saffron.* 47, 48

N. 5. *Variety of colours, in what flowers, from what Off-sets.* 49

CHAP. III.
Of Propagation by Stems, Cuttings or Slips.

N. 1. *A Catalogue of Plants this way propagable.* 50

N. 2. *Explication of the manner of propagation by stems cut off from the Mother-plant, or slipt, by Examples and Rules for particular direction.* 51

N. 3. *Experiments made of the success of the cuttings off divers Plants set in Water.* 53

N. 4. *The manner of growing by cuttings.* 55

N. 5. *Of propagation by the sowing small and almost insensible parts of Vegetables.* 56

CHAP.

CONTENTS.

CHAP. IV.
Of Propagation by laying.

Num. 1. *What Plants are this way increased.*   57
N. 2. *The example of this manner of propagation.*   ib.
N. 3. *Requisites for the manner of laying.*   58
N. 4. *Of propagation by Circumposition.*   59
N. 5. *Of the manner of growth by circumposition, and whether thence an argument may be made for the descension of Sap.*   60

CHAP. V.
Of Insitions.

N. 1. *Of grafting in general, and particularly of shoulder-grafting, Whip-grafting, Grafting in the cleft, and Ablactation; shewing the manner of doing these several operations.*   61
N. 2. *What Plants take on different kindes, with divers Experiments and Stories on this subject.*   66
N. 3. *Rules for Grafting and of Inoculation*   68
N. 4. *Kirkers Experiments concerning Insitions examined, and opposed by new Experiments.*   73
N. 5. *The manner of growing by Grafts, Historically set down, with addition of some Philosophical considerations.*   74

CHAP. VI.
Of the ways for, and Seasons of setting Plants.

Num. 1. *Of cultivated Plants.*   79
N. 2. *Of the setting of Woods, Fruit-Trees, and Plants uncultivated.*   81
N. 3. *Whether any Vegetables may be set so as to grow in the Air.*   84

CONTENTS.

CHAP. VII.

Of the means for the Improvement and best culture of Corn, Grass, and other Vegetables belonging to Husbandry; and of the ways for removing the several annoyances that usually hinder such advantage.

Num. 1. *Of the annoyances to Land, and the Impediments that usually distemper it, to the disadvantage of the Husbandman.* 86

N. 2. *Of the remedies proper to cure the excessive coldness and moisture in Lands, and the ways of improvement thereby, in grounds subject to these distempers, by dreining, Pigeons and Poultrey dung, Urine, Soot, Ashes, Horse and Sheep dung: O Ground cold and dry, and how these Soyls may be applyable thereto.* 87

N. 3. *The ways of improvement of dry, light, sandy, gravelly, flinty Land, by floating, Marl, Chalk, Lime.* 92

Num. 4. *Remedies for accidental annoyances and hindrances of Improvement, particularly the ways to destroy Fern, Heath, Ant-hills, Moss, Rushes, Rest-harrow, Broom, or any such Weed or Shrubs, that infest the ground: Whether liming of Corn prevents blasting, the effects of that and Brine in Improvement: Concerning Moles, and the ways to destroy them or drown them; a way of Antipathy, as to this effect, in Animals and Vegetables to the Bodies of their own kinde, when they are in the way of corruption: of the change of Seed; and Mr. Blith's way of preserving Corn from Crows, Rooks, &c.* 97

CHAP.

CONTENTS.

CHAP. VIII.

Of the Means of Improvement and best culture of of such Plants or Flowers as are usually cultivated in Gardens or Orchards, and of the ways used for the removing such annoyances as are commonly incident to them.

Num. 1. *Of the annoyances in general incident to Garden Plants.* 103

Num. 2. *Of defences of choice Plants from cold.* 104

N. 3. *Of shade requisite to sundry Plants, especially when young, for their defence from the Sun and Winde; and of watering, necessary to cultivated Plants.* 107, 108

N. 4. 1. *Examples of the best Culture of Hops, and ways of ordering them after they are first set, taken out of Mr.* Blith. 109

2. *Mr.* Parkinsons *way of ordering the Seedlings of Tulips grown.* 111

N. 5. *Of annoyances by Plants growing too thick and near together, and of the remedy thereof, and improvement by pruning Trees, and setting them at great distances; plucking off the young Germens of Garden-flowers, to make the rest more fair; of the sizing of Turneps, Carrots, Parsneps; of Weeding.* 114

N. 6. *Of Pismires, Earwigs, Canker and rottenness in choice Plants, Catterpillars, Mossiness, Bark-binding, Bursting of Gilly-flowers.* 122

Num. 7. *Of improvement and melioration of divers Sallad Herbs, by blanching or whiting, from the French Gardiner, and Mr.* P's *Observations.* 125

Num.

CONTENTS.

N. 7. *Of Acceleration and Retardation of Plants, in respect to their Germination and maturity.* 129

Num. 8. *Of melioration by Richness, or other convenient Minera in the Soyl, for the feeding and better nourishment of several Plants: Of artificial Bogs, and the change of Seed, as a means to bring fair Flowers: Of Exossation of Fruit, or making it grow without Stones.* 134

N. 9. *The conclusion of the Treatise, with one or two choice observations of the wise and good Providence of God, which may be seen in the admirable make of Vegetables, and fitness to their ends, which are not generally taken notice of, but are, with many more, overseen by men busie in the affairs of the world.* 139

ERRATA.

Page 8. col. 2. l. 17, r. *Scorzonera.* p. 10. l. 5. are but young p. 11. l. 15. *properata satio,* p. 27. l. 6,8,9. r. Serotine, p. 33. l. 21. r. foe that to bear Seed yearly, is general to all, unless p. 61. l. 3. and I am well contented, p. 94. l. 22. as possible p. 117. l. 20. adapted p. 149. l. 18. Vestments.

THE HISTORY OF
Artificial propagation of Plants.

CAP. I.
Of Propagation by Seed.

Num. 1. *Of Propagation of Vegetables in general, with a Preface to the Discourse.*

He Illustrious and Renowned Lord *Bacon*, in his Discourse concerning the advancement of Learning, reckons it among the Deficients of Natural History, *That the Co-operation of Man, with Nature in particulars, hath not been observed; and that in those Collections which are made of Agriculture, and other manual Arts, there is commonly a neglect and rejection of Experiments, familiar and vulgar, which yet to the interpretation of Nature,* and which I shall adde, general profit, *do as much, if not more conduce, then Experiments of a higher quality.* The same noble Person, in his par-

partition of Philosophy, complains of the want of an Inventary of what in any subjects by Nature and Art is certainly, and may be undoubtedly wrought. I believe his Lordship hath had many of his minde in former, has now, and is likely to have in future ages; for amongst those few Writings extant on these Subjects, some prove altogether useless, as being so full of their natural Magick and Romantick Stories, that we know no more what to credit in those Relations, in the Natural, then what in civil History we may believe of King *Arthur*; *Guy* of *Warwick* in ours; or of *Hector* and *Priam* in the *Trojan* Story: Others elevated in their Fancies, write in a Language of their own, addressing their Discourse to the Sons of Art, speaking rather to amuse, than instruct, and prove like blazing Stars, that distract many, and direct few.

Many of those who would write for Universal Instruction, either know the things that might make up the matter of their History, but want the skill to draw up such an Inventary, as his Lordship requires, as common Tradesmen and Artisans; or else indeed are learned enough to draw up the writing, but stand aloof from the knowledge of most of the particulars therein to be ingrost; which is the ordinary case of us, such of us as have pretensions to Scholarship.

I being necessitated by my obligations and respect to a Person truly Noble, to give some account of the particular effects of Man, co-operating with nature, in the matter of our English Vegetables, as they are improved by Husbandmen and Gardiners, desire to undertake no more, but to give a sincere endeavour, That the way of the Artist be set down, and the effect of Nature thereon; in the first of which, I intend

tend my directions so plain, as if appointed for the instruction of some Artists rude and untaught Apprentice: and the second's, if not so homely, yet as easie and evident, being a little disgusted with any thing intended for the use of Philosophy, when overgarnished with Rhetorical Tropes, which like Flowers stuck in a Window for whatsoever intended (either cheat or ornament) certainly create a darkness in the place. *Behemenical, Paracelsian,* and such Phrase as many Alchimists use, I must for the same reason avoid.

In the drawing up the Inventary, I will study that it may be true in all parts, and not to mingle, according to the example of *Pliny, Weeker, Porta,* and many more, both Latine and English Writers, any false relation, without its distinguishing Character; and if it be not perfect, it shall be for want of skill, or present remembrance of particulars.

The end of the Artist is to Propagate and Improve: To propagate, is to multiply the individuals of each kinde: And to improve, is to bring them, being propagated, to a more then ordinary excellency and goodness. The ways of increasing the particulars of each kinde, are, 1. By Seed, 2. By off-set, taken from a Mother-Plant. 3. By laying the Branch of a growing Plant down into the Earth. 4. By bearing up a Soil to it. 5. By Stems set without roots. And lastly, By the various ways of grafting and insitions.

Concerning all these, as likewise the preservation and melioration of things propagated, I shall endeavor to enumerate what Plants may be increased by each of these ways, and to shew how the operation in each may be performed, and what the product is that

that by nature thence ordinarily enſues : Definitions are hopeleſs in this matter, uſeleſs too, and it might be harmful : If I ſhould define Sowing, to be the caſting of Seed into the Earth, in ſuch maner, and at ſuch time, when in the ſurface of the bed the earth would ſo ferment, as might be proper to the explication and further germination of the Seed and increaſe of the Plant, there might a world of controverſies ariſe about the particulars therein contained ; and yet all that is there would be uſeleſs, till the particular Plants, and the maner of the operation, and time required to the ſowing of their Seeds be firſt declared : I ſhall therefore wave all ſuch endeavors, and haſten to what may rather prove for uſe than pomp.

N. 2. *A Catalogue of Plants that may be encreaſed by Seeds.*

Aconite.
F. Adonis.
Alliſſanders.
Alkanet.
Alaternus.
Alliaria.
Almonds, the bitter from our Engliſh Fruit, ſerving for his own kinde, or to make ſtocks for Aprecots and Peaches.
Ammi.
Amaranthus.
Angelica.
Anemones.

Aprecots.
Aparine.
Apple-Trees of all ſorts.
Apples of Love.
Arſemart.
Armerias.
Archangels.
Ariſtolochia.
Aſh.
Aſparagus.
Aſphodels.
Avens of all ſorts.
Balm Apple.
Balſamina.
Baſil.

Balm.

Balm.
Barberies.
Bay-Trees.
Beech.
Beans.
Bears-ears.
Betony.
Bell-flowers.
Beets.
Bistort.
Bitter Almonds.
Blite.
Blew-bottle.
Bloodwort.
Bryonies.
Bulbous Violets.
Burrage.
Buglosse.
Burdock.
Burnet Saxafrage.
Burnet.
Burrs.
Buckthorn.
Bullets of all sorts.
Cabbage Plants.
Campions.
Carnations.
Calamint.
Camomile.
Caucalis.
Carrots wilde.
Carrots.
Caraway.
Carduus Benedictus.

Centory
Celandine.
Chickweeds.
Chondrillas.
Chervil.
Cherries.
Chesnuts.
The Cornelian Cherry.
Cichory.
Citrulls.
Ciches.
Claries.
Coleworts.
The Seed of Clematis, but it comes not up till the second year.
Coleflower.
Corn of all sorts.
Coronopus Ruellii.
Comfrey.
Corianders.
Columbines.
Convolvulus major, minor, and other Bind weeds.
Cornsallet.
Coronopus.
Most sorts of Cowslips.
Crown Imperial.
Cranes-Bills.
Crowfoot of most sorts.
Cucumbers.
Cumin.
Cyclamens.
Cypres from out-landish seed.

seed.
Dandelion.
Dames Violet.
Some Daysies.
Diers Weed.
Dittander.
Divels bit.
Dittany.
Dill.
Docks.
Dogs-bane.
Earth-nut.
Egrimony.
Elecampane.
Endive.
Epatica's.
Eupatorium cannabinum.
Evergreen Privet.
Ewe.
Feverfew.
Fennel flowers.
Fennel.
Fenugreek.
Figwort.
Fig-trees.
Fibberds.
The Firre-Tree.
Some Flags.
Flowers-de-Luce.
Flos Adonis.
Flaxes.
Fleabane.
Fluellens.
Foxgloves.

Frittelaries.
French Mallows.
Fumitery.
Garlick.
Garden cresses.
Germanders.
Ginny.
Gilly-flowers.
Gourds.
Most of our English Grass; to this end, Husbandmen use Hay-dust (as they call it, in which lie the Seeds of their grass) to sow upon such Grounds as they mean to turn from Fallow into Pasture, or where they would have the Grass grow thicker.
Grain of all sorts.
Groundsel.
Groundpine.
Gromwell.
Hawkweeds.
Hartwort.
Hawthorn.
Haselnuts.
Henbane.
Hemp.
Hellebores.
Hercules his all heal.
Hyacinths.
Horse-radish.

Horned

Horned-Poppy.
Hony-wort.
Horehounds.
Hounds Tougues.
Holyoke.
Honyfuckles.
Holly or Holme.
Hypericum.
all Hyſſopes.
Indian Pepper.
Ironworte.
Juniper.
Kidney-beans.
Knapweed.
Knot-graſſe.
Ladyſmocks.
Lamb-lettuce.
Lark-spurs.
Lavander.
Langdebeefe.
Leeks.
Some *Lillyes*, though but few.
Lychnis Calcedonica.
Linum umbellatum.
Lovage.
Lupines.
Marjoranes of all kinds.
Mandrakes.
Maſtique.
Common Marygolds.
Mallows.
French and African Marigolds.

Marſhmallowes.
Maſterwort.
Maple.
Malacotones.
Melons.
Melilot, and its kinds.
Medlars.
Mercuries.
Molyes.
Motherworte.
Muſtard.
Muſcipula.
Mulleines.
Mulberries by seed from hotter climates than our own; for our heat ripens not the seed.
Mirtles likewise.
Narciſſes.
Dead-Nettles.
Stinging Nettles.
Noli-me-tangere.
Nightſhades.
Nigella.
Oke.
Onions.
Some of the *Orchis* of ſtones.
Orach.
Orpines.
Paronychia.
Pancies.
Peucedanum.
Parſley.

Parſnip

Parsnips.
Panax Herculeus.
Pellitory.
Pennyworts.
Peonyes.
Pease.
Pease everlasting.
Pears.
Peaches.
Periclimenum.
Pinks.
Pimpernel.
The *Pitch-tree.*
Plums.
Plantains.
Wild and garden *Poppyes.*
Pondweed.
Pompions.
Primroses.
Ever green *Privet.*
Pulsatillas.
Purslane.
Quinces.
Radish.
Ragworte.
Rampions.
Radix-cava.
Reeds.
Ribwort.
Rosemary by Out-landish seed, sometimes by our own.
Roman Nettles.
Some *Roses*, the Flower being not gathered, but left to seed.
Rocket.
Rushes of many sorts.
Rue of all sorts.
Some of the *Saffrons*, and *Mede Saffrons*, whose seed lyes under the earth.
Satyrions.
Savory.
Sabina baccifera.
Scorpion grasses.
Scurvey grasse.
Scorodonia.
Scabiouse.
Scorzoneca, but it comes up with some difficulty.
Seseli æthiopicum, or *Hart-wort.*
Sesamoides,
Shepheards purse.
Skirrets.
Sloes.
Smalladge.
Sneezewort.
Snapdragon.
Sowthistle.
Sorrels.
Spiderwort.
Spinach.
Spurges of many kinds.
Spignel.
Stitchwort.

Starr*d*

Starreflowers.
Stockgilliflowers.
Starrewort.
Flowers of the Sun.
Sword-flags.
Swine-cresse.
Swallow-wort.
Sycamores.
Tarragon.
Teasels.
Terra-glandes.
Thorney Apples.
Thorough-wax.
Thyme, both the Winter and Summer sort.
Thistles.
Tabacco.
Thlaspies.
Toad-flaxes.
Tragopogon.
Trefoile, and its kinds.
Tulips.
Turnips, and all its wilde kinds.
Tutsan.
Venus Looking-glasse.
Vervain.
Vetches.
Violets.
Vipers-grasse.
Virgine-bower.
Umbilicus-Veneris.
Vines from outlandish feed.
Water-betony.
Water-lilly.
Wallnuts.
Winter-cresse.
Winter-cherries.
Willow-weeds.
Woolfs-bane.
Wormwood.
Woodroof.
Wood-sorrel.
Woad.

There is a great controversie concerning *Harts-tongue, Maydenhair* of divers sorts, *Scolopendrium, Fernes,* and other Plants, whose property is to have the back of the leaf lined with a brown dusty substance, whether this be a seed, or onely particular mole, and character of Plants of that nature.

I dare not disbelieve this, when perfectly ripe, to be a true seed, because divers, very experienced persons (as Mr. *Bobart* particularly) affirm, that they have seen the small Plants, or Seedlings at a distance all

all round the Mother-plant grow up as is ordinary from shed seed of other plants, and by *Miscroscopes*, the likenesse of this dutt to other seeds is apparently seen.

N. 3. *The Seasons of Sowing.*

First, the most naturall time of Sowing is that which Nature it self followes (*viz.*) when the seeds of their own accord fall into the ground.

At this season may be sowen all stony seeds that can endure the Winter, as Cherries, Plums, Peaches, Apples, Peares, likewise all Nuts, Buckthorne, Ash, Oke, and most wild English Plants, though they may as well be sowed any time before the Spring.

The seed of hot, and sweet hearbs, as Thyme, Savory, Marjerome of some kinds, and other hot hearbs, if they get any reasonable strength and growth before the frosts, doe well enough; also Angelica seed, Scurvey-grasse, and the seed of Bears-ears, Aniseed, Fritellary, Crocus; and, for ought I know, all the rest of Bulbous-rooted flowers: So Tulips and Anemones thrive best, and come soonest, being sowed after the seeds are gathered, or in Autumn: For many *October* does well, but care must be had to keep tender Plants from Frosts and the violence of Winter weather, when they but young from the seedlings. If you doubt the nature of any seed, divide your quantity, and sow some of it in the Spring, some before the Winter.

At this time also must be sowed divers Plants, for that by experience 'tis found, that being sowed in the Spring they will not grow at least not that year:

Of this kinde Myrrhis, or sweet Chervill, and all Rubarbs, which easily grow then, but faile being sowen in the Spring.

The mistake of the time has made some admire, that when they with care had sowen Angelica seeds severall times together, this never grew; on the contrary, the Seed being shed would grow in any place, never so uncouth or stony; nay even carried away by the water, would grow wherever it was lodged in the banks, and that well and lustily; whereas the reason of the difference was in the season, for the laborious Artist kept the seeds till Spring was his hindrance, whereas better instructed Nature would have committed them to the earth many months sooner. 'Tis a true Proverb, *preperata satio solet sape decipere, sera semper.*

Some seeds are sowen at the breaking of the Frost, and the very first beginning of Spring, and that upon a hot bed, for the greater security and speed of the Plant to be propagated: So the early Radish, the Sensitive Plant, Maracoc, Balm Apples, French Marygolds, Muskmelons, all Cucumbers, African Marygolds, the Marvail of the world, the Indian Cresse, or yellow Larksheel, Lettices that they may be had early.

The hot Bed is made with horse-dung laid four, five, or six foot high, and of the same breadth commonly, increasing or diminishing the quantity of the dung (which uses to be fresh, as it comes from the stable, mingled with stale Litter, Hay, &c.) according as you would have the heat greater or lesse, upon which bed of dung you lay fine mould, five fingers breadth in deepnesse or thereabouts, compassing it round with hay-bands which keep the dung together,

ther, and hinder the steaming out of the heat by the sides; then staking it up with stakes, and putting bended sticks in the manner of a very low roofe to hold up tilts that are put to secure the Plants, the hot bed is perfectly finished. Those that use Cap-glasses, or Casements made to lye upon a frame over their beds, neverthelesse must use, though not tilts, yet covering with straw, litter, or the like.

Asparagus and Chervil are best sown in Winter before Christmasse, or shortly after, and in the beginning of Spring without any hot bed. In *February*, or afterwards, are sown Parsnips, Leeks, Onions, Aniseeds, Carrets, Radish, Spinage, Larks-spurs, Marygolds, Cærefolium, Corn-sallet, and with the first of these the Rounseval pease.

Colliflowers and Cabbages in the middle of *February*, Muskmelons somewhat after, or then for a venture. 'Tis observ'd by all I have enquired of, that the lesse of the Winter the Cabbage or Collyflowers feels, the more subject 'tis to Caterpillars. In *March* or *April* (or according to some with us, from the beginning of *February*; or, if the Frosts breake, any time in *January*) Carrot, Radish, Tobacco, Fennel, Cresses, Skirrets are ordinarily sowen.

In *April*, Marjerome, Basil, Coleflowers; for by often transplanting and care you may have Coleflowers from seed, sowen in the Spring, though it be very far gone even to *June* or *July* the same yeare, Pincks, Armeriaes, Convolvulus, Kidney-beans, Lupins, Hyssope, Lavander, Stock-gillyflowers, Thyme, Hemp.

About the latter end of *April*, Purslane, Clove-gilliflowers, Carnations, Basil, Rosemary.

About

About *Midsummer* sow the early Pease, to be ripe six weeks after *Michaelmasse*.

Note that our Gardiners, though there be some perill, chuse to sow early, because they have much advantage by all sorts of forward commodities; so Turnips sowed early, many run to seed, yet one good then, is worth three at another season. The same may be said of Pease and Carrets, which by cold are spoyled many times; yet it is observed by some, that oftentimes, whether by difference of ground, or other accident, the Bean latter sowed wil overtake the former, and so in some sorts of Pease.

Many seeds are best sowen about *August*, so Turnips, and the black Radish, for a peculiar reason; which is, being sowen sooner, they are apt to run up to seed before Winter, and not to fil the root at all. Onions for winter provision, Lettice and Corn-sallet for the same occasion; Spinage too, alwaies upon that account, though otherwise they may be sowen with the first. Nay, our Gardiners here in *Oxford* sow Turnips in *April*, and so forward till the Winter.

Cabbage plants are sowed commonly about *August*; and the first Coleflowers, that they may before Winter be so grown, as to be transplanted at greater distance, so to abide till the Spring. I have known some, when frost has spoyl'd the winter Cabbage-plants, to have furnished themselves from plants raised in the Spring upon a hot bed.

Many seeds must be gathered a little before they are throughly ripe with the stalkes on which they grow; for should it abide until the full maturitie in the Garden, by wind and weather great part of the seed

seed would be shed, which will easily perfect its ripeneſſe as it lyes cut upon its ſtalk, being laid any where within doore upon a cloath or mat where the Sunne comes. Of this kind is Lettice, and moſt of thoſe ſeeds that ariſe from the flock with a wooli-neſſe.

There are many Plants that will grow in all times of ſowing, and therefore are ſowen many months, one after another; ſo Radiſhes, and Spinach, and Peaſe, which are ſowen with the firſt in the Spring; and ſo month after month till Autumn. Thoſe Lettice which abide the winter are wont to be tranſplanted to Cabbage in the Spring, even as Cabbages are with admirable ſucceſs.

Our Gardiners, that they may have Cucumbers to ſel one under another, plant them in hot beds from *February* even till *May*.

Peaſe are ſowen from the beginning of *November* (or by ſome a fortnight before, though with ſome danger of the biting froſt) and ſo forward til after *Shrovetide*.

Rounſevals, if ſowed never ſo early, will ſcarce come before the latter part of the Month of *June*.

Husbandmen generally uſe to ſow Wheat under furrow in the Autumne; but I have ſeen it with good ſucceſs ſowen in the Spring, and harrowed in after the manner of ſowing Barley; the crop being as good as any other times upon the ſame ground, after the uſual country procedure.

Some ſeeds muſt be ſowen dry, not after raine or watering: Of this kind is Myrrhis ſeed, Baſil, Scorzonera, and all ſuch as being wet run to a Muſcilage.

Many times they ſow divers ſeeds in a Bed together,

ther, as Radishes and Carrots, that by such time as the Carrots come up, the Radishes may be gone. Upon beds newly set with Licorice they sow Onions or Radish, or Lettice if their Licorice plants or ground be but weak, so as not quickly to cause a shadow with their leaves. *London* Gardiners sow Radish, Lettice, Parsley, Carrots, on the same bed, gathering each in their seasons, and leaving the Parsnips till the Winter; before which time they are not esteemed good, or wholsome.

Note, that where your grounds are very warm by reason of hedges, hot beds, dunghils, &c. that may abate the power of the frost, seeds may be ventured into the ground much sooner than otherwise in ordinary places.

Cabbage seeds and Coleflowers are sowed in *August*, or so timely as to be exactly well rooted plants before winter; and this is the best way: Or are sowed after, so that they are transplanted in the time of cold. This way is hazardous in the winter, by reason of the nipping Frosts, and chargeable, in that they require much attendance, and covering, and uncovering, which those plants that are confirmed before winter doe not. Secondly, they are more subject to Caterpillars in the Summer; but the way of raising of them by hot beds in the Spring for Cabbages is the worst way of all, and most subject to the peril of that vermine.

Those Plants of the Spring sowing, that you sow later than ordinary, require to be the more watered and shadowed from the heat.

Those in the Spring that are sowed earlyer than ordinary, require the more to be defended from the cold.

Those in the Autumne, that you prematurely sow, are to be watered and shadowed the more. Those which you sow late are to be better defended from the Winter till they have gotten strength.

N. 4. *Examples of Sowing with some particular directions for some choice Vegetables.*

Examp. 1. *From* Mr. Parkinson; *directing skillfully the ordering of* Tulips *in their propagation by seed.*

The first example I shall give you out of Mr. *Parkinson*: The time (sayes he) and manner of Sowing Tulip-seed is thus, you may not sow them in the Spring of the year, if you hope to have any good of them, but in the Autumne, or presently after they be through ripe and dry; yet if you sow them not untill the end of *Octob.* they will come forward never the worse, but rather the better: for it is often seen, that over-early sowing causeth them to spring out of the ground over-early, so that if a sharp spring chance to follow, it may goe near to spoile all, or most of the seed: We usually sow the same years seed, yet if you chance to keep of your own, or have of others, such seed as is two years old, they will thrive and doe well enough; Especially if they were ripe and well gathered: you must not sow them too thick, for so doing hath lost many a Peck of seed; for if the seed lie one upon another, that it hath not roome upon the sprouting to enter or take root in the earth, it perisheth by and by; Some use to tread down the ground where they mean to sow their seed, and having sowen them thereon, doe cover them over the thicknesse of a mans Thumb, with fine sifted earth, and they think they
doe

doe well, and have good reason for it: For considering the nature of young Tulip roots is to runne down deeper into the ground, every year more then other, they think to hinder their quick descent by the fastness of the ground, that so they may increase the better. This way may please some, but I doe not use it, nor can find the reason sufficient; for they doe not consider that the stifness of the earth doth cause the roots of the young Tulips to be long before they grow great, in that the stiffe ground doth more hinder the well thriving of the Roots then a loose doth: and although the roots doe runne down deeper in a loose earth, yet they may easily by transplanting be holpen and rais'd up high enough. I have also seen some Tulips not once removed from their sowing to their flowering; but if you will not loose them you must take them up while their leaf or stalk be fresh and not withered: for if you doe not follow the stalk down to the root, be it never so deep you will leave them behind you.

The ground also must be respected, for the finer, softer and richer the mould is, wherein you sow the seed, the greater shall be your increase and variety. Sift it therefore from stones and rubbish, and let it be either fat naturall ground of it self, or being muckt, let it be throughly rotten: some I know to mend their ground doe make such a mixture of grounds, that they mar it in the making.

Ferrarius bids that the seed be sowen in *Septemb.* (as soon as rain shall make the ground fit) half a fingers breadth in good Garden mould, not to be removed in two years after, at which time they are to be removed and placed in severall beds, according to their seveall bignefs, where in 4 or 5 years they will bear their flowers. *Example*

Example 2, Of Anemone's

Within a moneth after the seed of Anemone's is gather'd and prepared, (in *August*, faies *Ferrarius*, or three dayes before the full Moon in *Septemb.*) it must be sown, for by that means you shall gain a year in the growing, over that you should doe if you sowed it the next spring : If there remain any Woolinefs in the feed, pull it afunder as well as you can, and then fow your feed reasonably thinne upon a plain fmooth bed of fine earth, or rather in pots or tubs, and after the fowing fift or gently ftrew over them fome fine good fresh mould, about one fingers thicknefs at the moft for the firft time; and about a month after their fpringing up, fift, or ftrew over them in like manner (this is a neceffary circumftance) another fingers thicknefs of fine earth, and in the mean time if the weather prove dry, you muft water them gently and often, and thus doing you fhall have them fpring up before winter and grow pretty ftrong, able to abide the fharp winter, in their Nonage, in ufing fome little care to cover them loofly with Fearne, furze, or Bean-ftraw or any fuch things, which muft neither ly clofe to, nor too farre from them.

The next Spring after the fowing, or which is better the next *August* you may remove them, and fet them in order by Rowes with fufficient diftance one from another, where they may abide, until you fee what manner of flower they will bear.

Many of them being thus ordered, if your mould be fine, loofe and fresh, not ftony, clayifh, or from a middin, will bear flowers the fecond year after the fowing, and moft of all of them the third year, if your

ground

ground be free from smoaks and other annoyances. Nay Mr. *Auflen* of *Wadham Coll.* a skillfull florist, assured me that he has had Anemones from the seed sowed in summer, that were in flower within ten moneths of the time of their sowing.

N. 3. *Clovergraffe being esteemed as great an improvement as any our ground is capable of: I shall adde such speciall directions as are given for the ordering thereof: Sir Richard Westons observations and rules are as followeth.*

Clovergraffe-seed thrives best when you sow it in the worst and barrenest ground. Such as our worst heath ground in England. The ground is thus prepar'd for seed.

First pare off the heath; then make the paring into little hills: you may put to one hill as much paring as comes off from a Rod or Pole of ground, which is the square of sixteen feet and a half. The hill being sufficiently made and prepared (as they doe in *Devonshiring* as we call it) are to be fired and burnt into ashes. And unto the ashes of every hill you must put a peck of unflake Lime; the Lime is to be covered over with the ashes; and so to stand til Rain comes and flakes the lime. After that mingle your ashes and Lime together, and so spread it over your land. This done; either against, or shortly after rain, plough and sowe; ploughing not above foure inches deep and not in furrowes; but as plain as you can, and to make it yet plainer, harrow afterwards, and that with bushes under your Harrowes.

The ground being thus prepared you may sow your seeds. An Acre of ground will take about ten pounds of Clover-graſſe-ſeed, which is in meaſure ſomwhat more then half a Peck. The chief ſeaſon for ſowing it is *April* or the latter end of *March*.

About the fift of *June* it will be ready to be cut. It yeelds excellent hay. The time of cutting it will be more exactly knowne, by obſerving when it begins to knot: for that is the time: And ere the year be done, it will yeild you three of thoſe crops, all of them very good hay; and after you have thus cut it the third time, you may then feed the ground with Cattle all the winter, as you doe other ground.

But if you intend to preſerve ſeed, then muſt you expect but two crops that year, and you muſt cut the firſt according to the foreſaid directions, but the ſecond growth muſt be let ſtand, till the ſeed of it be come to a full and dead ripeneſſe, and then muſt you cut it, and threſh the tops, and ſo preſerve the ſeed, you ſhall have at leaſt five buſhells of ſeed from every Acre.

This ſeed thus threſhed off, there will be left long ſtalks, theſe your Cattle will eat; but when they grow old and hard, you are to boile thoſe ſtalks and make a maſh of them, and it will be very nouriſhing either for Hogs, or any thing that eat thereof.

After the ſecond cutting for ſeed, you muſt cut that year no more; but as it ſprings again, feed it with Cattle. One Acre of it will feed you as many Cowes as ſix ordinary Acres, and you will find your milke much richer; which induces ſome not to cut it at all, but onely to graze it for their Dayry.

Being once ſowed, it will laſt five years, and then being plowed, it will yeild three or four years together rich crops of wheat, and after that a crop of Oats.

And

And as the Oats begin to come up, then sow it with Clover-seed (which is in it self excellent Manure) for you need not bestow any new dressing upon the ground, and by that time you have cut your Oates, you will find a delicate grasse grown up underneath, upon which if you please, you may graze with Cattle or Horse all that year after, and the next year take your crop as before at pleasure.

To prevent mistake, I must give this advertisement, that whereas Sir *Richard Weston* commends heathy ground, he is not to be understood, of such dry and barren ground without its best Manure by chalk, lime, and the like artifices of husbandry. For otherwise it has failed in the growth & improvement thereby expected. Mr. *Blith* commends ground naturally good, betwixt ten and twenty shillings an Acre: giving this generall Rule, that no land can be too good for Clover that is not too good for Corn.

Hempe and Flax are used to have the same culture, and the best husbandry that I have observed of them has been in *Staffordshire*, where this procedure is generally observed. About the beginning or middle of *Aprill* the flax seed is sowen upon new broken ground, immediately upon its being broken up. The seed they either have from their own Crop, or buy it from a warmer Country: Mr. *Blith* reports the true East-Country seed to be farre the best, who for tryall of both, sowed on the same land, the Ridge or Middle with our Country seed, and both the furrowes, with Dutch or east-country seed, (such as is bought in the seedsmens shops at *Billingsgate* in *London*) the effect was that our seed, though on the ridge it had the advantage of the ground, was encompassed with the Dutch, as with a wall about it, so much the Easterne

seed

seed did out grow it. He likewise for warmer parts, as Essex and Kent thinks *mid-March* a convenient season for sowing it: If weeds grow therein they carefully weed their crop and pull it in dry weather when it lookes yellow, lest growing over ripe it blacken and mildew, and tye it up in handfulls that it may perfectly dry. Then they ripple it, is, that they get out the seeds by drawing it through an Ingine like an iron double tooth combe, which they call a Ripple: the boles of seed pulled off, they lay on a boorded or playsterd floore to dry, it being dryed they lay it up and thresh it not out of the boles till *March*, when they winnow it clean from the huskes.

The watering of it is thus: The Flax being well dryed, they bind up about 20 handfulls in a bundle and putting many of these bundles together they stake them down in the water, that they may not be carryed away by the Streame. The flax abides in water 4 or 5 dayes and nights, then they spread it on the grasse that it may dry, turning it every 3 dayes, and when it is full dryed they lay it up and house it, and when they see their occasion they use their Brake and Crack, instruments devised for the purpose to bring the Tow from the Flax. The whole Government and husbandry of hemp from the seed to the distaffe is so like this of Flax that the same example and rule may very well serve for both.

Woad, according to Mr. *Bliths* directions, is best sowed where you sow your Barly or Oates, upon that very husbandry or tilth, about the middle of *March*, and may grow up among the Corn because it groweth not fast the first summer, but after the Corn is cut it must be preserved; it requires a rich and warme soil.

This

This plant is of great use to Dyars, and coloureth the bright yellow or lemon colour, It abates the strength and superrichnesse of land, and may prepare for Corn in land of its own Nature too rich, which is, as Mr. *Blith* observes, sometimes a fault, though not so frequently as the contrary extreme.

Beans require a low deep ground and Waterish, not dry, sandy or gravelly soyle: this is true of feild beanes, though I first tooke notice of the great difference in our London Gardens, where the labourers for their own eating would give one part in three more for a measure of beanes from the former than from the latter soyl, who assured me that from the same seed and care garden beans have much more meale, pulpe, or kernell and thinner skins in the moist than in the dryer and lesse waterish ground.

N. 4. *The Generall observations for the manner of sowing.*

Besides the Examples aforesaid, I shall adde some rules such as by Gardiners are usually observed.

This is generall that all seeds must be covered with the earth, which is done, either by sowing the ground and turning the seed in under the furrow, or by drawing trenches in the soyle, and then drawing the earth over them with a hoe, or sowing the beds ready drest, and hacking in the seed with the same instrument, or by harrowing, raking with a rake or drawing bushes over the sowed ground to cover the seed, or to set the single seeds with a stick, or lastly to sow the ground and afterwards to sift or strow fine mould thereon.

The two laste wayes are for choice seeds when the workeman desires to loose none for want of burying the sowing under furrow is for such seeds as must endure the winter, the depth of ground being part of their security against the winter colds: nor are all seeds of strength to shoot their germen through so much earth. The sowing intrenches is used for Pease, there being thereby spaces left between the rowes, of half a yard more or lesse, to gather them as they ripen, and roome whence to draw mould to the roots, which frequently done, is very advantageous to them. It is likewise handsome for Spinach, Endive, Thyme, Savory or other garden herbs to grow in rowes after this manner of sowing.

Moisture is absolutely necessary for the growth of all plants, two or three dayes after a great rain is accounted a good season; in dry weather two dayes after rain say the London Gardiners, agreably to that of *Ferrarius*, *Nec tamen simulac magnis imbribus terra permaduit seres, sed tantisper expectabis, dum pluvius ille mador modice exsiccetur, ne madenti limosoq; in solo statuta radices exputrescant* de Fl. cult. l. 3. c. 1. Seeds that are apt to run to a Muscilage are unfit to endure moisture upon that account, as elsewhere I noted.

I prescribe nothing concerning the observation of the faces of the moone, because I much doubt of any effect therefrom. Neither doe Gardiners that work, nor Authors that write, prescribe alike rules; but contradict each other in their direction, for the particular observation of this Planet, as to any intended production. Nor is it agreeable to my reason, that the moones being in the full at the first explication of the two dissimilar leaves, or germination of the plant, should cause a double flower, this germination according to this present History, differing little from other augmentations of the same plant, in opposite qaurters quarters immediately enfuing: fo that if a full moone be proper, I fee no reafon why it may not be effectuall, by vertue of the fame phafis the third, as the firft or the twelveth, as the fixt day of the feedlings augmentation.

The meliorating of ground belongs to the head of Improvement; here I fhall only obferve that where ground is very light, as in fome London and Kentifh gardens, it is found profitable after fowing to tread in the feed.

Some fteep all garden feed before they fow them to make the germination the more fpeedy, but feeing there be no better wayes of infufion than in Earth and Water, why the fame bofome of a well watered ground fhould not be moft fit for this operation I fee not.

In feeds that are long in coming up, the feed bed is not to be digged up the firft winter: For I know diverfe feeds that will for a great part of them ly under ground the firft, year and come up the fecond: of this Nature is the Afh-key fometimes, the Peach, Malecotone and fome Plums.

N. 5. *Of variety of kindes, different in colour, tafte, fmell, and other fenfible qualities, proceeding from fome feeds, and what plants they are that bring feeds yeild-fuch variety.*

In Carnations you have feeds that give admirable Variety from the Orange-tawny Carnation and all his ftrip't kinds that are double and keepe their tawny in them in any meafure. The white, Tawny and Carnations darkly fpotted, *Ferrarius* commends for producing variety of colours and ftripes. Kernells of divers Apples and Peares bring variety of kinds, different in tafte, fmell, colour, and hardneffe, and are

are as often promoted to better, as the degenerate to worſt, as I am very credibly informed, by perſons that profeſſe themſelves to have ſeen the experience. The kernells of the Burgundy Pear has brought a noble alteration and produceth a pear farre beyond that excellent kind: Peaches and Malecotones doe ordinarily the like, ſo that by ſeed is thought to be their beſt propagation.

Our Gardiners in chooſing the ſeed of ſtock-Gylliflowers to make them bring double ſtocks, take their ſeed from ſuch tops as bring fine leaves in their flower, eſpecially if it be one ſtrip't; but Mr. *P.* ſayes thoſe that bear double ſeeds, cannot be diſtinguiſhed from the other, and I have reaſon to beleive him, for ſuch as chuſe their ſeed this way, doe not find that it anſwers their expectation.

For Tulips that are early, or Præcoces, the purple ſays Mr. *Parkinſon*, I have found to be the beſt, next thereto is the purple with white edges, and ſo likewiſe the red, with yellow edges; but each of them will bring moſt of their own Colours. For the Media's, take thoſe colours that are light, rather white then yellow, and purple, then red, yea white, not yellow, purple, not red: but theſe again to be ſpotted is the beſt, and the more the better; but withall or aboveall, in theſe reſpect the bottome of the flower (which in the precox Tulipa you cannot, becauſe you ſhall find no other ground in them but yellow) for if the flower be white or whitiſh, ſpotted, or edged and ſtraked, and the bottome blew or purple (which is found in the Holias, and in the Cloath of Silver, this is beyond all other the moſt excellent, and out of queſtion the choiſeſt of an hundred, to beget the greateſt and moſt pleaſant variety, and raritie, and so in degree the meaner in beauty you sow, the lesser shall your pleasure in varieties be: Bestow not your time in sowing red or yellow Tulip-seed, or the diverse mixtures of them, they will (as I have found by experience) seldome be worth your paines. The Serolina being not beautifull, brings forth no speciall varietye: *Ferrarius* lib. 3, chap. 7. commends the Serolina for seed, (but I find he makes but two sorts; Præcoces and Serolin's) and among them the white, with the black purple, or blew bottomes or Scarlet with skycoloured bottome inclining to purple; for both them will (sayes he) bring Tulips mark't with varietye and handsomnesse: But Tulips without a blackish bottome are noe good breeders of various coloured flowers.

The two lesser Spanish bastard Daffodills, the leaves of which are of a whitish green colour, one alittle broader then the other, and the flowers pure white, bending down their their heads, that they almost touch the Stalk again, give Seed from which springs much varietye, few or none keeping either colour or height with their mother plant.

The seeds of divers Sowbreads, by name the Roman Sowbred with round leaves, the Autumnall Ivy leaved Sowbread, some flowers-de-lis, and many sorts of Bears-eares doe the like in produceing admirable variety.

As for Anemones, take't from Mr. *P*. and our common dayly experience that there is not so great variety of double flowers raised from the seeds of thinne leave'd Anemones as from the broad leaved ones. Of the Latifolias, the double Orange-tawny seed being sowen, yeildeth pretty varietyes, but the purples, or reds, or crimsons, yeild small varietyes, but
such

such as draw nearest to their originall, although some be a little deeper or lighter then others: But the light colours are they that are chief for choice, as white, ash-colour, blush or Carnation, light Orange, Simple, or party-coloured, single (or double if they bear seed) which must be carefully gathered, and that not before it be fully ripe, which you shall know by the head, for when the seed with the woollinesse beginneth a little to rise of it self at the lower end, then must it be quickly gathered, left the wind carry it all away, after it is thus carefully gather'd it must be layd to dry for a week or more, which then being gently rubbed with a little dry sand, or earth will cause the seed to be better separated, although not throughly, from the woolinefs or downe that compasseth it.

In the seed of the Mervayle-of-the-world, take notice, that if you would have variable flowers, you must chuse out such flowers as be variable while they blow, that you may have their seed! for in this plant if the flower be of a single colour, the seed will likely bring the same.

N. 6. *Some other relations of transmutation, and the possibility of a change of ones species into another examined.*

I have often heard persons affirme, that they have sowed Barley, or some other grain, and in the ground the seed has been so altered as to send forth Oates instead of corn, according to its own species. I am as yet farre from giving any assent to this their History. The Reasons why I disbeleive them are, first, because the Relators affirme whole fields to be thus varied, and that to one species (viz) of Oates, which is different from Barley in the straw, eare and grain

it

it selfe. Whereas in the variation of seed, in those vegetables, in which the change is undoubted, the colour only or some other easily alterable accidents, such as the sensible qualities are generally found are transmuted, and this transmutation ends not at all in another divers kind; but in severall small diversities of the same kind; The storyes of Wheat turned to Mustard-seed were as likely to be true, and is a fit parallell to create a right beleife of the true cause of the mentioned effect. Secondly, I knew a Gentleman who plowed a piece of land in the spring, and then sowed it not, but after it was harrowed and prepared for seed left it to its own Genius and nature to produce what it was inclined to: The Ground was off its own Nature apt to bring forth wild-Oates amidst the Corn, now in defect of Corn there grew as many wild-Oates unmixt from any other weeds, as the land could carry. This was tryed in a great peice of land, and much proffit was made of the Oates, the Gentleman having cut them green for Fodder *Anno* 1657.

My judgment therefore is, That the fallacy which befell my above named Relators was, that they mistook the cause of the production of the Oates mentioned; for to me it is much more easie to conceive, that by some evill accident, as it often happens (the seedcorn being corrupted and perish't in the ground) the ground it's self from its own Seminary, sent out the suppositicious Crop of Oates or Mustard, than that there should be a variety of so strange a Nature, and declension from its property, in the issue of any species.

It is indeed growen to be a great question, whether the transmutation of a species be possible either in the vegetable, Animal, or Minerall kingdome. For the possibility

possibility of it in the vegetable: I have heard Mr. *Bobart* and his *Son* often report it, and proffer to make oath that the Crocus and Gladiolus, as likwise the Leucoium, and Hyacinths by a long standing without replanting have in his garden changed from one kind to the other: and for satisfaction about the curiosity in the presence of Mr. *Boyle* I tooke up some bulbs of the very numericall roots whereof the relation was made, though the alteration was perfected before, where we saw the diverse bulbs growing as it were on the same stoole, close together, but no bulb half of the one kind, and the other half of the other: But the change-time being past it was reason we should beleive the report of good artists in matters of their own faculty.

Mr. *Wrench* a skilfull, and industrious gardiner for fruit and kitching-plants told me that the last year there was a change betwixt the kinds of the Coleflower, and the cabbage. Others I know who as from their experience most confidently affirme that they have prime-roses of the milk white colour, the root whereof before in another ground bare Oxelips: and it is usually beleived that divers single flowers may be changed into double by frequent transplantations; made into better grounds. I knew those that have had the wood Anemonies, and Colchiums double, who affirme that they took them into their garden wild, and single, and that that change was made by the soyle, and culture of the place.

For the animall Kingdome the instances of transmutation are in silkwormes, cadiz, and all caterpillars, which after a long sleep from the reptile turne into the volatile kind.

The minerall Kingdom is supposed to be famous

and

and fruitfull in these changes, the hope of the Philosophers stone, or perfecting medicine requiring this beleife: Yet I am persuaded that in many of their changes they rather separate, and bring to apparence a latent minerall, than produce it by the transmutation of another into that nature. *Sennertus* recants those writtings of his, that affirmed iron to have been turned into copper by naturall, and artificiall waters of Vitrioll. The effect only in his second, and more mature judgment being the separation of a copper before latent in the Vitrioll, and the precipitation of it by the parts of the iron: and I have seen some experiments made by the honorable Person, for whom I am now writing, that have added strength to my former persuasion, particularly the supposed transmutation of quicksilver into lead, published as real by the learned *Vntzerus* and others, and to be made by dissolving the quicksilver in aqua fortis, & precipitating it by the tincture of Minium, proved but sophisticall, the Lead produced that way being indeed not made of the Mercury, but only reduced out of the tincture of Minium, wherein it lurck't, as that Gentleman doth more circumstantially set down in his own papers, and others of the like nature, which it were not proper here further to insist on.

It is a question, whether there be any reall transmutation, from the vegetable to the minerall kingdome, in petrifaction of any sort of wood: those petrifactions, which I have seen in England, are made thus, some particles of stone, that impregnate the body of water, make a crust about the stick that is to be petrified, and enter into the pores thereof, as fast as they are layed open by the water, washing through the stick, wherein there interceeds, noe change of

the

the same parts, but by addition of some, and subtraction of others, if I imagine aright, the new effect is wrought. The proof whereof may be, that the fibres of wood appear visible and to the touch and taste amidst the body of the stone.

In Ireland there is a Lake wherein (as that Noble Person I but now mentioned, hath related to me) there is soe great a petrifying faculty that the best whetstones used in that nation, are made of wood, cast therein to be petrified. In which stones though all the lineaments of the woody fibres remain, yet they are indued with the hardnesse, and other qualities of an exact stone. And Corall, the entire stonynesse thereof noe man can doubt, may well be imagined to be originally a vegetable bearing root, stalk, and leafe; and that afterward it is turned into its hardnesse by the peculiar property of the water: whether these operations of nature are likewise perfected by addition and subtraction of parts only, or whether it be required that some parts for the production of this effect be transmuted I shall not determine.

And for the deciding the whole question, if the form be specificall, and so made by the aggregation of a certain number of accidents, those accidents & that number must be assigned that are thought enough to compleat a new form, before we may begin to judge in this matter for that very many accidents maybe changed it appears by the above named instances in vegetables & in other bodyes many more: Vinegar and Wine, are the same parts transposed and yet there seemes to be more difference between them than between Endive and Cichory, Maidenhaire and Scolopendrium, Rubarb and Dockes, which are in Vegetables esteemed for diverse species formally or specifically distinguished. Numb.

N. 7. *Of Provision for seed.*

Many Rootes are to be transplanted at the latter end of the year, and will bring forth perfect seeds: as, Carrets, Parsneps, Turneps. Cabbages are to be layd in Cellars all winter, the roote and Cabbage being replanted in the spring, or the seed may be got, though not in so plentifull a manner, from the stalks of Cabbages, whence in the season the Cabbage was taken either replanted, or standing in their old places: Coleflowers give their seed from the like care that is bestowed on the Cabbage.

I have seen Gardiners that provide Cabbage-seed in great quantitie for the shops in London upon their courfe ground, to sow Cabbage seed which without transplantation shall bring forth Coleworts for boyling hearbs, and then a crop of seed: many plants that bear fruit bring their feed every year in their fruits, so Apples, Peares, Plumes, Peaches, Aprecots, Wheat, Barley, Rye, Peafe, Beanes, and many that beare no fruit doe the like, so Lettute, Radish all grasses, so that unlesse some peculiar plants which require to be excepted: Yucca Indica, bears neither flower nor seed in lesse than four years time: 'tis generall that each seed will ripen every year, and the best generall token of maturity is its loosneis from the pedall by which its joyned to the stock, so as kernells in ripe Apples grow loose from the core.

Those persons that make Verjuce or Cider can best furnish him that intends a Nursery, for notwithstanding both the violence of Mill or Presse, the kernells escape entire enough for Vegetation; but care must be had that they be immediately sowen after the pressing

D leaft

left being layd on a heap they heat, in the manner of wet Hay, and burn the germen of the seed, which in the moisture of the bruised fruit by that heat will prematurely sprout forth to its own perishing.

In providing Lettuce seed, mark the plants that you see stronger for seed, and after they have begun to shoot stalks, strip away the lowest leaves, for two or three hands breadth above the ground, that by them the stalk be not rotted.

Let Carnation and Gillyflower-Cods of seed stand upon the Roote so long as you may, for danger of frost, then cut the stems off with the Cods on them and dry them so, as not to loose the seeds; The drynesse of the Cods and blackness of the seed is an Argument of ripenesse: *Ferrarius* Lib: 3. Cap. 15. Reports, that the bottome of every Cod brings the best seed: and the largest flowers.

The seed of Crocus's are only, or at least, best taken from the ordinary stript vernall Crocus, the great purple Crocus, the great blew Crocus of Naples, the stript purple, the lesse purple, flame coloured, the purple with small leaves, the yellow stript, the cloath of Gold. Clovergrasse and seeds of that nature, are provided by letting the grasse run timely to seed, particularly by mowing it about *May* and thence abstaining till the seed is through ripe.

Such seeds as are weighty and sinke in water are best; the contrary are usually languid and unfit for propagation.

Out-Landish seeds are used for such plants, whose seeds cannot be got here for want of Maturity, or any other reason.

The Spanish-Muske-Melon-seed is accounted best, though we use our own with good successe: few Gardiners

diners here will use their own Onion-seed, for they find it runnes to Scallions: Myrtle with us comes not to seed, nor Mulbery. For the sensitive plant, the Amaracoc or Passion flower &c. we send for seed to the Barbado's.

What advantage our Nation might have by propagation of exotique plants by seed brought new from severall Countryes beyond the Seas, tis hard to ghesse that there would be advantage tis certain. I remember that *Bellonius* a man very diligent, and much employed about knowing the nature of plants, growing in other Countrys than his own, which was France, wrote a whole book to shew the possibility and advantage of this improvement, to perswade Merchants to furnish gentlemen with seed, and them to use it. Tis known that Peaches, Aprecots, Nectarins were lately not only strangers to England, but to France likewise. Mulbery is likewise an Exotique plant, and by King James his Command sent for over and propagated by seed.

Exotique Seeds are good not only to propagate plants yet not with us, but likewise to make a more plentifull production then can with ease be made from any other way. of propagation of such we already have.

Care must be had in sowing seed, or at least in setting them, where you intend that they shall thrive, that the ground bear the best proportion may be, to the places and the particular Minera of the places where such plants in other parts use to grow, not to put mountainous plants in low and moist grounds. Why the Taurick Cedars, were they planted in Walles, should not grow I know no reason.

It were worth the while to confider in all feeds, whether there be noe diftinguifhable difference in the feed, that may be of ufe, as to fooner, or greater growth. In the fame bed divers feeds being fowed of one kind, particularly Apples, Peares, Plummes, Cherrys, or Peaches, fome Apple feedlings will in the fame mould, and diftances, much outfhoot the reft of the fame kind, and fo in the Pears, and other kernels: it might here be enquired, whether the great or leffe, fend bigger plants, and of fpeedier growth ? as it is by fome obferved in buds, that the fairer the bud is upon the fheild and ftronger, the better thrives the inoculation, and not only growes more certainly but more luftily.

2. Whether the Canker in pippins, arife not from an incongruous grafting, and it were not better to bring them up from kernells, or graft them on a more mild ftock than that of a Crab.

Whether there might not be gotten diverfe years fooner trees of ftature from kernells of great bodyed and quick growing Apple trees, and fuch whofe kernells vary not much their kindes, than from Crabbs, which is a wood of a flow growth and harfh Nature.

N. 8. *The manner of growing by feed.*

The feed is confidered either as allready made, or as it is under the hands of Nature, imperfect, yet in the way to be made.

In it made, there are confiderable, firft, the Coates and cotton that cover it about, and preferve it from injuries; fecondly, the effentiall and proper parts of the feed it felf.

Many feeds have two Coates above the Cotton,

and one thinne one under, next invefting the feed, such are Sicamores.

All seeds that I know have within their Covers actually a Neb, which anfwers to a roote, which is joyned to leaves more or leffe in number: betwixt the ftalks of, or amidft thefe leaves there is a bud, eye or Germen, juft oppofite to the Neb, or initiall Roote, but by reafon of its fmallneffe it is fcarce difcernable in many feeds till it begins to fpring.

1. Moft plants have only two leaues actually joyned to the Neb, which are commonly very unlike the proper leaves of the plant: of this fort are the flowers of the Sunne, Ediffarum Clypeatum, Cucumbers, Melons, Amaranthus, Thiftles, Thlafpyes, Mallows of divers kinds, Arch-angells Spurges, Nettles, Clary, Orach, Dill, Parfely hath two leaves diffimilar, but not much foe, Melilot two diffimilar, and one, if I miftake not, fimilar.

2. Many plants have more Leaves in the arifing from the Neb, as Creffes have fix.

3. Some plants have but one diffimilar leafe as Anemones, Tulips, Fritellaryes and all bulbous fpring flowers that I have obferved. Wheat, Barley, Rye, all grain and graffes that I know have a germen wrapped up att one end of the grain in a hofe or fheath which germen confifts of leaves wrapped about the bud by a plica, or folding made the long way of the leafe, not overthwart as in Sicamores, Maples and other complicated leaves of feeds. Nor doth the whole corn divide it felfe into leaves, and coates or huske as in thofe examples, but the greater part thereof containes a meale which by the heat and moifture of the foyl is turned into a pappy fubftance not unlike the Chyle found in the lacteals of animal bodyes,

bodyes, and may be as I suppose, reposed nourishment for the young blade at such time as the earth would prove but a dry Nurse. I have taken notice that Carnations come up sometimes with three, sometimes with four leaves, though the most have but two: and it is Mr. *Bobarts* observation, that such as come up with more leaves than two, prove double flowers, which if it generally holds true, it were a compendious way to weed out all the rest at the first coming up, to avoid the labour of culture of such plants as in the end will not prove advantageous for profit or pleasure.

Beanes, Pease, Kidney-beanes, Lupines, have this peculiarity, that the grain being clett, each half is as one of these dissimular leaves, which is usually contained in every seed, and between these thick leaves are contayned other similar leaves, or such as differ but in growth or bignesse from the true leaves of the Plant. 'Tis to be observed in all these great seeds, that though the pulse, or thick part of the grain perish, yet if the Neb and smal leaves are entire, the seed may prosper; as I have seen Feild-beanes that have been eaten through with wormes, prove good thriveing seed. But tis reported, that Pismires have learned the wit to spoyl the seed from growing in their store-houses, by biting off the very Neb before they repose the grain.

The growth of the plant from the seed is thus; by convenient moysture and heat, the Neb stricks through the Covers, and goes directly down, if not impeded, in earth or water, a convenient way, ordinarily, two or three inches, in which time the leaves either rowled up, or otherwise inclosed, break their bonds, and explicate themselves, being lifted commonly a little higher by the growth of the stalk, or lengthned

neb;

Neb: and you may obferve, that the growth above ground, at the firft motion upward, is nothing proportionable to the motion downward. After the root is well made and fafhned betwixt the leaves that were actually contained in the feed, there arifes into more plain fight and appearance, that little Germen before, in many plants fcarce feen, like to that bud, which is left on plants in winter, which fpringing, brings forth the true leaves and Branche, of the plant fowen.

If I am enquired of, whether each feed has a compleat effence and diftinct form of its own. Nay further, whether it be a true and perfect plant? I muft fay that I have found it fo to be, even more than an egge, a liveing thing, and immediately nourifhable It has root to grow, body to bear the port of the plant, Bark to direct the Sap into all its parts, and germen or bud to fecure the meanes of future growth, and to boote leaves, which is all and fomewhat more than in the winter the fturdieft Oke can boaft of.

It has been accounted an Intereft in Philofohpy hererofore, and that in our Schooles, that feed fhould not be efteemed an actual and formal plant, becaufe of divers abfurdities, that if feed were animall, would happen in their Schoole doctrine; as that there would be pluralities of formes in the fame trees; The Soule might be divifible into parts; The fame thing might be agent and patient; Nay fome have faid, that it may be of dangerous confequence in Divinity, if it were granted, that feeds had the actuall formes and effence of that thing whofe Seeds they were.

I am glad tis noe Herefy now, to appeale to fenfe from a Doctors opinion, and that I may freely in this matter require to be tryed by my garden, though it be againft the fentence and Judgment of the Doctors

Conimbra, *Suarez*, *Ruvio*, *Pererius*, *Bonamicus*, *Fonseca*; and that we begin to lay aside the fear, that from a certain truth, ill consequences may arise: That Canon will certainly hold longest which is best built in the bottome.

It is conceived by some that the immediate cause of the Growth of the seed, is the Spirit working upon the Salt and Sulphur, Earth and other constituent parts or Elements of the Seed: For the Spirit is supposed to be made Volatile by the heat of the earth and water, which in Spring and Autumne, (the cheife times of germination) is of a proper temperature for fermentation; and then the spirit being so Volatized, and riseing up and expanding it self every way augments the whole plant, and distends the sides of the seed, whereby the growth of the seed plant is effected.

But how it comes to passe, that the conveiance of these expanded particles is ordered to proceed, according to the lineaments of each Vegetable, noe person to my knowledge has yet made any conceit; and it being beyond any ocular discovery of the most acute Searchers, to finde out the Conduits or Trunckes serving to so intricate a carriage, and how it comes to passe, that a seed first, has its Neb thrust downe without dilatation of the sides, and then, how the upper part of the Neb or germen orderly frames the Vegetable above ground in so trim a body, rather then a confused masse, I take it not for any part of my taske to enquire.

I shall likewise leave it to the imaginations of Philosophers to determine, whether upon the distention made, it be by an elective faculty in the Seedling, filled up with similar parts drawn from the Earth, and to by Nature originally fitted specifically for that plant: or whether there being a continual motion of particles from the earth, pressing upon the plant, those only get entrance whose shapes and figures are such, as correspond to the pores in the young Vegetable; which meeting in the body of the plant with its constituent parts in nature not unlike themselves, they easily are joyned thereto, and so cause an augmentation in the whole: or whether dissimilar parts, either to fill up the Vacuum made by distention, or for other reasons, got up into the plant, doe obtain there a change of nature, and from the form, Soul, Archeus, or other principle, are altered from their first being, into a likenesse of nature with the Seedling, and become homogeneous to it; These are Questions, in the determination of which, till I am better informed, I desire to take no side.

N. 9. *Of the cause of Greennefs in the leaves of Vegetables.*

It has been made a question by some what it is that causes greennesse in all Herbes, especially such whose feed, and the stalk, and Leafe, contained therein are white, and whether the cold beating of aire and water upon Vegetables may not have some influence in the production of this effect.

I truely have been tempted to think the affirmative, which is that the coldnesse and brisknesse of the free aire, in plants that grow in the land, and the like qualitie of the water, in water plants produces the verdure or greennesse, that is generally the beauteous Vestment of all Vegetables, or at the least has some considerable influence as to this production: for by experience

perience I have proved that plants being in a close roome, brought up from seeds in pots, or otherwise, the leaves and stalks prove to be white, or pale, & not green, which is according to the Lord *Bacons* experiment, who *Cent.* 5. *Exp.* 47. setting a Standard Damask-Rose-Tree &c. in an earthen pan of water, where bearing leaves in the winter, in a chamber where no fire was, the leaves were found (as his Lordship relates) more pale and light coloured, then leaves use to be abroad; which palenesse, I suppose to be greater or lesse, proportionably to the freshnesse and freenesse of the aire that the plant enjoyes. Grasse will likewise change its colour, if by any weighty body, or other lying upon it in the field, it be kept from the aire: The truth is, all plants have peculiar delight in the aire, which I have proved by this Experiment; I have taken young seedlings in a pot, and put them in a window where there was a quarry out, the seedling would immediately leave its upright growth, and direct its body straight to the hole, and so become almost flat and levell with the earth in the pot: Then turning the pot so, that the inclination of the stalk might be from the hole, the plant has then crook't it self in form of a horn, or the letter C. to the aire again. Upon the Second turn of the pot, the upper horn being placed from the aire, the plant would, with its upper part, return to the open place, and leave the stalk now in the form of an S. Nay, sometimes I have bid persons tell me, which way they would have such a plant grow; they have marked the place in the brime of the Pot, that mark I have turned to the hole in the window, by which means the plant without any force, and that in not many houres space, hath inclined its stalkes to the mark made.

That

That the aire has great influence in producing the verdure of plants, may likewise not improbably be argued from the Experiments of Blanching, or whiting the leaves of Artichockes, Endive, Mirrhis, Cichory, Alexanders, and other plants; which is done by warm keeping of them, without the approach or sentiment of the Coole and fresh aire; whereby all plants that otherwise would bear a green colour, become exactly white.

Hence it may likewise bee, that the roots of most Vegetables that are under ground, and covered from the aire, are white generally, whereas the stem, and upper parts of them are ordinarily green, and many rootes that are by nature of a peculiar colour, as Radishes, yet the point of the roote that is deepest in the ground, retaines a whitenesse, as well as other roots, being in that part of the roote removed from the aire, the red part commonly standing above or just in the surface of the earth.

Hence also it may be, that those leaves of Cabbages & Lettuce that are expanded in the free aire are green, those that being covered with their felloves : and secluded from the blasts of wind and weather, and kept in a warme Covert, become as white as any thing that is artificially blanch't.

True it is that, there be plants that grow in the bottome of waters, and so cannot be supposed to have this help from the aire, otherwise than as the aire chills the water, and the water having received this qualitie from the aire, makes the like impression upon its domestique plants.

CHAP.

Chap. 2.

Of Propagation by offsets.

N. 1. A Catalogue of Plants which may be propagated by offsets and suckers arising with Roots from the stool and Roote of the Mother Plant.

Aconite or *Wolfes-bane*.
Adders-tongue.
Alexanders.
Anemones.
Angelica.
Aristolochias.
Artichockes.
Asphodels.
Asarum.
Asparagus.
Avens.
Barberies.
Barrenworth.
Bawme.
Bears-eares.
water and wood *Betony*.
Bistort.
Spanish *Broome*.
Butchers Broome.
Brooklime.
Briony.
Barts, and such like Apples.
Buglosse.

Burdocke.
Burnet.
Calamus aromaticus, which requires moisture.
Camomill.
Caltha or *March Marigold*.
Cherryes where the stock is not grafted.
Chives.
Cinquefoyle.
Clownes all-heal.
Costmary.
Cowslips.
Comfrey.
Cowslips of Jerusalem.
Coltsfoote.
Columbines.
The *Crown* imperiall.
Crowfoot.
Cuckowpints.
Dames violet.

Daysyes

Dayſyes.
Dens Leonis bulboſus.
Dittander.
Docksteoth.
Dockes.
Dorias his wound wort.
Dragons.
Dulcamara, or woody night-ſhade.
Egrimony.
Elmes.
Elicampane.
Everlaſting Vetch.
Ewe.
Fernes.
Feverfew.
Figtrees.
Filbeards.
Filipendula.
Flowers-de-Luce.
Fleuellen or Speedwell.
Galingall.
Garliques
Gentianella.
Germander.
Gooſberryes.
Golden-rod.
Ground Jvy.
Haſelnuts.
Harts tongue.
Herba paris.
Helleborine.
Hellebores.
Hercules all heal.

Hyacinths.
Horſradiſh.
Houſeleeke.
Horſemints.
Hops.
Horſetaile.
Jaſmine.
Jeruſalem Artichoke.
Kentiſh Codlings.
Knapweed.
Lovage.
Lady's bed ſtraw.
Lilyes.
Lilium convallium.
Lunaria.
Lungwoort.
Mandrakes, for often there may be take from them particles of their roots, which will grow well, though the uſuall way of their propagation is by feed.
Marſhmallowes.
Maſterwort.
Madder.
Mints.
Moly.
Monkshood.
Mulberryes.
Mugwort.
Nurſe-gardens.
All ſorts of Orchis, or Docks-ſtone.
Petaſitis.
Periwinck

Periwincle.
Peony.
Pease.
Pilewort.
Poplars.
Potatoes.
Prunella.
Primroses.
Pulsatillas.
Raspes.
Radix cava.
Reeds.
Roses of most kindes.
Ruscus or Butchers broome.
Rubarbs.
Satyrions.
Saponoria.
Sanicle.
Scabious.
Sedum.
Serpillum.
Setfoyle.
Skirrets though seeds will produce better.
Smallage.
Sorrells.
Solidago Saracenica.
Solomons Seal.
Some Spurges.
Stitchwort.
Strawberryes.
Sword flags.
Tarragon.
Tansey.
Thistles.
All sorts of Tulips.
Valerians.
Some Vetches.
Vervaine.
Times.
Violets except the yellow.
Water mints.
Water Lillyes and most of the other water plants.
Winter Cherryes.
Willow weeds.
Woolfes bane.
Wormewood.
Yarrow.

N. 2. *The way of making Offsets by Art.*

Nature usually provides this help of propagation, without the wit or industry of men, called to her assistance, but that not generally in all plants, nor alwayes in any one : and therefore I esteeme it well deserving any mans learning who delight in Gardens, to know any meanes to enlarge this way of propagation beyond

beyond the bounds it is carryed to by natures course. There is a pretty way (which in truth I first learned from Mr. *Bobart* our Physique Gardiner) for the making Offsets where nature never intended them; which is done by bareing the root of plants of woody substance, and then making a cut of the same fashion with that which is made in laying: Into this cleft a stone must be put, or something that will make the root gape, then cover the roote over three inches with mould, and the lip that is lifted up will sprout into branches, the roote of the old tree nourishing it. When the branches are growen, cut off this plant with its Roote to live of its self.

If you can, leave an eye on the lip of your roote, which you after the incition lift up; for the branches will then more speedily and certainly issue out of the root so cut.

In Bulbous Rootes, *Ferrarius* makes offsets thus: If (sayes he) a Bulbous root is barren of Offsets: either put it in better earth, or cut it upon the bottom in the crown of the roote whence the fibres spring, and that but lightly with your naile, and sprinkle some dryduft as a medicine to the wound; and the effect he affirms to be this, that so many wounds as you shall make, into so many offsets shall the genitall vertue dispose it self.

N. 3. *Rules for direction in taking off Suckers, or Offsets.*

Care must be had, that the Damme be not destroyed in her delivery from her new brood, which may easily be done, if too great a wound be made upon the stoole, or mother-plant, by tearing off the Suckers. Tis *Ferrarius* his peculiar precept about Anemonyes

monyes: That they be sure as to take off such Offsets that will scarce hang on, so not to teare off such as hold fast to the mother-plant, for that would be to the peril both of the offset and motherplant. Yet I have seene the very substance of Sowbreads to have beene divided with a knife through the heart, and yet grow well on either part, when they have not afterward been over glutted with wet. Flaggs-Beares eares, Primroses and Cowslips, and generally all rootes, that are not Bulbous or tuberous must have, and doe require a violent separation, but the lesse the wound is, the better shall your plant thrive, and be lesse subject to corrupt by the moisture in the earth.

In the replantation there is required the generall care of young sets, all plants of fibrous rootes are assured in their growth, by convenient watering, but for bulbous and tuberous the Gardiners hand is, and ought to be more sparing, because that moisture is a peculiar enemy to these plants, and often rots them, if it get into any crany of their rootes.

N. 4. *Examples of planting by Offsets.*

Licorice requires the richest & most forced ground, very deep, that there may be roome for the downright roote, light, without stones or gravell, and dry from moisture: The sets are made either from the runners that creep along the upper part of the ground from the roote, or else are taken from the Crown of the master-roote, and are set at a foot distance or lesse in *February*, or *March*, according to custome, though I suppose any time in the winter might as well serve the turn, the richer the ground is, the further they may be set apart.

Hoppes

Hoppes require to be planted in a very rich well foyled land, and not moorish, unlesse the bog be first well dreyned, the stronger the setts are, the more immediately will proffit arise from the Garden, if three or four inches about, they are so much the better, let the center of the hills be ten foot removed each from other, that so you may put the more poles on a hill, and both the sun and plow may have free passage between them: those that have lesse ground make lesse distances, and toyle their garden with the spade, and put but three poles to a hill, whereas such as plant 9 or 10 foote distance, use four at the least, if not five: In planting, which is thought to be best done, when the frosts are past, (some prescribe *April* for the season) there is nothing required but that they be set about the center of the place, intended for the Hill upon the plain surface of the ground in good mould, about three, four, or five in number according to the bignesse of the Hill intended, and ordered with the usuall care of offsets: besides this particular that as the sets grow the hill must be raised to their heads.

Saffron delights in a reasonable good and dry light ground, not extreamly foyled or moist, 'tis planted cheifly in some parts of Essex, Suffolke, and between that and Cambridge, at Saffron-Walden. They are set in the manner of bulbous roots, being taken when the bulbe is at the fullest, commonly about Midsummer, the bulbs are set by a line, (that the beds may be weeded with a hoe) and that either with a setting stick or by trenches made in the manner of those wherein garden pease are usually sowed. This beares in the middle of the flower three chives, which is the Saffron, to be gathered every morning early and dryed for use, every second or third year at the furthest the beds must be replanted, and the offsets drawn away.

The

The generall way of this propagation is to take the offsets that rise from the bulbous and tuberous rooted plants, as Tulips, Anemones, Narcisses, Crocus's, &c. & the suckers which from the roots of poplars, Elmes, Nuttrees, Peares, Burts, Nursgardens, Kentish Collings, Gooseberryes, Roses, Ruscus, Calamus Aromaticus are very plentifully are drawn, and more, or less from all mentioned in the Catalogue. N. 1. Chap. 2. and to replant them in the seasons of setting, which are related in the proper chapter for that operation, into proper beds, and in convenient distances for their future education and growth.

N. 3 *Variety of colours, in what flowers, from what offsets.*

Our Gardiners respect most the roots of widdowes, for that they find by experience that they multiply the variety of Tulips not only from seeds, but from the offsets of these widdows: I my self have seen admirable declensions of them from their naturall purple and white.

The royall Crocus striped gives now and then very pretty variety from its offsets, as sometimes I have seen on the same roote an ordinary striped Crocus and another of a perfect flame colour, though the variety here be not so great as in Tulips.

Concerning the manner of growth by Offsets there is little to be spoken particularly, their roots being actually made while they remaine upon the mother plant, and their growth being like that of other well rooted vegetables.

CHAP.

CHAP. 3.

Of propagations by stemmes, cuttings or slippes.

N. 1. A Catalogue of plants this way propagable.

Abrotonum Vnguentarium.
Balsamita.
Barberyes.
Basil.
Basilmint.
Bay.
Baume.
Box.
Brooklime.
Burrs. and generally all such plants as break out into protuberances like warts upon the bark.
Bugle.
Cornelian Cherry.
Many Crowfooes.
Donas his woodwort being cut off neer the roote.
Elder.
Evergreen-Privet.
Germanders.
Gilliflowers.
Hyssope.
Jasmine.
Kentish Codlings.
Knotgrasse.
Lavander.
Lawrell.
Marjerome.
Marsh-mallowes. being taken up neer the roote.
Mastique.
Mulberyes.
Nursgardens.
Penny-royall.
Periwincle.
Pincks.
Polium monstanum.
Prunella or Selfe heale.
Quinces.
Some Roses, as the evergreen Rose.
Rosemary.
Rue.
Sage; both English and French.
Savory.
Savin. in moist ground and shadowy
Scordium.
Southernwood.

Southernwood.
Spearmints.
Strawberies, and generally all plants that have joynts upon creeping strings.
Thime.
Tripolium.

Veronica erecta.
Vines.
Violets.
Wall flowers.
Watercresse in water.
Withy.
Willow.
Woodbine.

N. 2. *Explication of the Manner of propagation by stemmes cut off from the Mother-plant, or slip't by example and Rules for particular direction.*

For example, I shall chuse to instance in Gilliflowers or Carnations, for which flowers observe this order, Seeke out from the stemmes such shoots onely as are reasonable strong, but yet young and not either too small or slender, or having any second shoots from the joynts of them, or run up into a spindle, cut these slips off from the stem or roote with a knife either close to the maine branch, if it be short, or leaving a joynt or two behind, if it be long enough, at which it may shoote anew: when you have cut off your slips you may either set them by and by, or else (as the best Gardiners use to doe) cast them into a tub of water for a day or two, then in a bed of rich and fine mould, first cutting off your slip close at the joynt, and having cut away the lowest leaves close to the stalke, and the uppermost even at the top, with a litle stick, make a little hole in the earth, and put your slip therein so deep that the upper leafe may be wholly above the ground (some use to cleave the stalk in the middle, and put a little earth or clay or chickweed, which we more use, within the cleft, this is Mr. *Hills* way in Sir *Hugh Plat*;
but

but many good and skilful Gardiners doe not use it; then close the ground unto the stemme of the plant.

As for the time, If you slip and set them in *September*, as many use to doe, or yet in *August*, as some may think will doe well, yet (unlesse they be the most ordinary sorts which are likely to grow at any time and in any place) the most of them, if not all, will either assuredly perish or never prosper well: the season indeed is from the beginning of *May* to the middle of *June* at furthest.

Ferrarius Lib. 2. c. 15. sayes, that from the moneth of *February* to the middle of *March* (viz) in the time of their germination, is the best time to slip this flower. He neither will have them slipt, nor twisted in the Roote, nor Barly put under them to raise adulterous fibres, but only advises that they be cut off in a joynt. The truth is, both the Spring and Autumne are good Seasons for makeing out Roots, the latter requires that the slip be so early set as that they may have time enough to take Roote, before the coldness of winter: The former, that the plant set in the spring, may have taken Roote before the Sun rises to emit violent and parching heats, which are generall Rules for Vernall and Autumnall settings.

Woody plants that bear leaves must be taken off, & planted some time between the fall of the leafe and the spring, some preferre the planting them in the beginning, some at the going out of the winter about the beginning of *February*, Immediately when the great frosts breake, at the first towardnesse to spring is a good season according to generall beleife.

Exam.

Experiments made of the succcesse of the cuttings off divers plants set in water.

Because in some disquisitions of naturall Philosophy, there may some matter of argument arise from experiments of the conversion of water into nutriment and substance of various and very different plants, whereof some are hot, others cold, some esteemed of a fresh, others of a salt nature, some in regard of mans body of healing, others of excoriating and blistring qualityes, some specifiques for the head, and the diseases thereof, others for the heart, and others for the wombe: I shall set down the truth of some few trialls concerning the growth or corruption of such cuttings, of divers Vegetables as without roots I kept in my chamber, in Vialls of water. Not willing thence to make any motion towards the restauration of the ancient doctrine concerning the production of all things out of water, or to rake up the scatter'd judgments of the once renowned *Thales*, which he made from the observation of the generation of fishes, and petrifaction by this element; as likewise from the influence (for he was aware thereof) and causalitie it has in the production and norishment of vegetable, and (if not immediately) by consequence of animall bodyes. Not desiring to make from these experiments (though I beleive the instance may be as well proper as specious) any argument for the more fashionable opinion of *Epicurus*, by shewing the various productions that may be made by the divers shufflings and positions of that which has the repute of the most pure and defecated element, but clearly intending to keep to my task, which is History, and rather to serve, than to be the Philosopher: I in short

rather

give the Reader this account: That *May* 1658, in Glasses of water the plants following grew from cuttings, and made themselves roots in the water, by name, they were Balsamita minor, Mints, Sedum multifudum, Penny-royall, Bugle, Prunella, water cresse, Purple-grasse, Periwincle, Dorias his wound-wort, Crow-foot, Brooklime, Marsh-mallows, Lawrell, Scordium, Tripolium, Knot-grasse, Nummularia, minima, Basil-Mint, Curl-mint, Horse-mint, Panax-coloni, Feverfew, and some others which I kept no account of, I have had at other times.

Plants that upon triall made by cuttings *May* 1658, did not grow being placed in Vialls of water were Mugwort, Rosemary, Stock-gilly-flowers, Alaternus, Lavander-cotten, Sage, Armeria's, Camomill, Rosemary, Polium montanum.

Stock-gilly-flowers, Bavne, Tansy, Groundsel, Lavander-cotton, Sage, Majorane, being likewise set in glasses of water dissolved into a muscilage, and so corrupted before they attained to any roots.

Plants that were corrupted by the water in some part of the stems and so dyed after leaves sent forth and roots shot, were, Basil, Mint, Marshmallows after it had grown a span, Panax-coloni, Palsamita minor, after six weeks growing, which made me doubt whether there were not the same reason of the dying of these plants that there is of grafts of Pears upon Apples, or Apples upon thorns, which grow for a while, it may be some years but surely dye before they arrive to any Maturitye: and secondly whether this reason was not the unlikenesse and diversity of parts between the stock to be nourished, and the nourishment apposed thereunto, for though some dyed after leafe and growth made, as purplewort parti-

cularly

ticularly by running into a Muscilage; yet generally there appeared noe such evident cause of their failing.

Plants that increased in weight, being planted in the water, were these, and the quantitie thus much.

Sedum multifidum in a moneth increased in weight, half a Scruple; Scordium as much in a fortnight. Donas his woundwort, grew in 6 weekes, gr. 13. Bugula in some what lesse time gr. 15. Watercresse gr. 25 in a moneth. Ranunculus half a Scruple in 6 weekes, and Periwinckle as much. Prunella, Brooklime, Scordium, and most of the sorts of mints got weight proportionably.

N. 4. *The manner of growing by cuttings.*

Such who desire to observe the working of Bees, get Casements to their Hives, that their eyes may not suffer impediment from the darknes of the place, for prevention of the same hinderance the use of beds of a Diaphanous soyl, in as Diaphanous bounds, or plainly of water in a glasse, I have found a proper remedy: and shall therefore from my observation of the growth of these particulars desire the reader will imagine the rest, or judge them alike, as truly so; what I remember I have always found them.

For the manner of plants growing by water, I observed that those plants that had many joynts easily grew and put forth roots only just at the joynt. Knotgrasse, Crow-foot, Panax-Coloni, all sorts of Mints, Penny-royall, Scordium, Bugle, Brooklime, Perivincle, which I conceive to be the reason why in setting them the practice is to cut of the plant just in a joynt, for so the roots immediately
spring

spring thence and no part of the stem corrupts, which it would, if it were cut of at greater distance.

In those herbs where there were no exact joynts, the roots sprung forth under some buds, as in Tripolium, Donas his woundwort, Marshmallows.

Every root that was made came forth first very white and single, but afterward in very handsome order and proportions, from thence arose other fibres striking every way in the water, where the side of the Vialls made no impediment to the growth of the spurtes issueing from the first and originall root.

N. 5. *Of propagation by the sowing small and almost insensible parts of Vegetables.*

Tis a generall received truth from common experience, that if the water wherein mushrooms have been steeped or washed, be powred forth upon an old hot bed, or the parts and offalls of Mushroms broken to peices bee strawed thereon, that from these parts as from a seed, there will speedily arise store of Mushroomes, every small particle of that imperfect plant being rather beleived seminall in the same manner as the boughs of Quinces &c. than that as in Adianthum, and diverse fernes, nature has disguised any particular seed clancularly to be the mean of Propagation in it.

Kircher the Jesuit affirmes that if you take an herb and shred it small, or reduce it into Ashes, these being sowed an herb will spring thence of the same species with the Ashes or shreds sowen: I thought that newes upon my first reading was too good to be true, and upon tryall made in very many sorts, could never make this way of propagation hold effectuall to the producing of any plant, and if it were

true

true it were an ill Cuſtome the Gardiners uſe to ſow their ſeeds with a great quantity of aſhes which are made from the wood or ſtraw and leaves of Vegetables generally and a wonder that they never ſhould come up amidſt the ſeeds moſt ſeaſonably ſowed.

Chap. 4.

Of Propagation by laying.

N. 1. *What plants are this way encreaſed.*

The plants that are uſually propagated this way, are Vines, Woodbines, Jaſmines, Mulberies, Savin: Evergreen privet in Woods all ſorts of Willowes and Sallowes to fill up bare places, Carnations, Gilliflowers, roſes, Horſ-cheſnut and all thoſe plants that will grow by Cuttings will this way grow with much more eaſe, by care and good watering gardiners doe apply this way with profit to ſuch plants as cannot well by any other meanes be encreaſed for want of ſeeds and offſets, and by reaſon of the repugnancy of their nature to grow either by cuttings or inſition.

2. *The example of this manner of Propagation.*

The moſt uſuall flower to be laid in Gardens, is the Gilliflower which every Gardiner here uſes, and is thus performed; Take thoſe ſlips you intend to lay, and cut the ſtalk juſt under that joynt of the ſlip, which is next the roote or middle ſtem, or under the 2d joynt half way through the ſtalk: then ſlit it upward to the next joynt from that under which you made your firſt inciſion, and put the top of a Carnation-leafe, or any other thing to hold open the ſlit, though

(though that be not altogether so needfull, for the cut being made on the lower side, and the Slip being towards the root bent down gently, as the manner is and the top of the slip raised with mould, the slip will be open of its own accord and remain so if you place it well) at the first some peg down the middle of the slip with sticks, that it may not rise from the positure in which tis first lay'd, you must remember to put good earth, enough to mould up your new Nursery, and to water it upon all occasions, and then in 7 or 8 weekes you may expect Rootes.

3. *Requisites for the manner of laying.*

1. To Laying, tis profitable if not necessary, that you (in the season of doeing this operation) cut the thing you lay, much in the manner you cut Gillyflowers, in laying them, unlesse in some plants that take any way as Vines, and 'tis so much the better if in Roses and other Layers of a woody substance, with an Awle you prick the stock at the place laid, as it is done in propagation by Circumposition.

2. Another Requisite is, that dureing the time of drought they be continually watered, and kept moist, otherwise they will make no exact roots perchance only a kind of knob or button full of fresh sap upon the tongue of the cut in the branch layd down, yet I have found these branches cut off with watering in the summer to grow well enough after their transplantation.

3. The seasons most fit for this operation, is, in the beginning of the spring or declension of the torrid heat of summer, that they may enjoy the moistnesse of such seasons most proper for the enticeing

forth

forth of roots, and most safe from excessive heat or cold.

N. 4 *Of propagation by Circumposition.*

Circumposition is a kind of laying, the difference is, that in this the mould is born up to the bough which is to be taken off: in laying the bough is to be depressed into the mould. Wee use this most in Apples after this manner, first break the bough a little above the place where tis separated from the main stock or arm, so that the hat or other Vessell that holds up the Mould to the incision or disbarked place may rest upon the stock, then slit an hat, an old boot, or take any strong peice of old course cloath, tying or sowing it so strongly that it may be able to hold up the mould to the incision, sometime before you fill this cap with mould, remember with an awle or point of a penknife, to bore two rowes of holes upon the upside of the cut about half an inch or more, one from another, then fill it with good mould, or such as is agreeable to the tree you work upon, and in the heat of summer, water it now and then, The time of this operation is not in the summer, as Mr. P. supposes (which mistake was sufficient cause why he should not like the experiment) but in the spring before the sap rises, particularly in *Febr.* or the beginning of *March*.

Such plants are propagable this way that might take by laying, but that the branches are too farre risen from the ground to be laid along therein; and therefore it becomes necessary, since they cannot stoope to the earth, that the earth should be lifted up to them.

N. 5. *Of the manner of growth by Circumposition, and whether thence an argument may be made for the descention of Sap.*

Concerning the manner of growth by Circumposition I shall only make this remark, whereas it is supposed by some, that the roots are made above the disbarked place, by the descention of the sap, which is supposed to be at the fall of the leaf, I have found experience very contradictory to their supposalls; for the leaves fall not till after Michaelmasse; and nature proceeds to the germination, and encrease of roots from the spring all the summer long, so that nothing can be argued rightly from this operation, or from the effect and product of nature thereupon for that opinion, which makes the sap to be every winter reposed in the roote, as in a large receptacle, and of its descention thither after every Autumne. If it were there as in a repository, it were a wonder that roots should be drier in *Decemb.* then in *May*, or *June*, and sensibly more devoid of juice. And if it did descend after Autumne, how could it ascend at the same time? That it doth then ascend is plain from this experiment; Take up a tree, or other vegetable, in the fall of the leafe; the leaves will wither, and the bark begin in a little time to wrinkle; then set it again in a proper soile, well watered; the effect will be that the leaves will recover freshnesse, and the bark wax plump and the body frime, and full as before, which could not be but by a fresh supply of ascending sap, which might fill up the pores made by the weather, and exhalation of the sunne. I am contented to beleive that the sap is in winter where I see it to be, (viz) on the body of the tree coagulated, or crusted into a new coate, encompassing the whole, which was not extant

the

the year before, and on the top fashioned into new shuits which visibly appear the product of that matter the place of which is asserted to be elsewhere and not I am as well contented not to suppose it abideing where upon the most sedulous inquest it cannot be found.

Chap. 5

Of Insitions:

N. 1. *Of Grafting in generall and particularly of shoulder-grafting, Whippe-Grafting, Grafting in the cleft and Ablactation.*

Grafting is an Art of so placeing, the Cyon upon a stock that the sap may passe from the stock to the Cyon without impediment. For the right operation of which it is a cheif remarke, that the space which is between the bark and the stock is the great Channell for conveiance and keeping of sap; so that every one that grafts well so orders the manner, that these spaces be so laid that the passage may be easy and direct from the spice under the bark of the stock, to the spice under the bark of the Cyon

This may be done severall ways.

First by shoulder grafting, the operation of which Mr. *Austin* do's well describe thus: Cut off the top of the stock in some smooth streight place that may answerable to the streightness of the graft when set on; then prepare the graft thus, observe which side is straightest at the bottome, or bigest end, so that it may

fit

fit the ſtraight part of the graft when ſet on, then cut one ſide only of the graft downe aſlope about an inch long or litle more, and cut through the bark at the top of the cut place: and make it like a ſhoulder, that it may reſt juſt upon the top of the ſtock, but cut not this ſhoulder to deep, (only through the bark or litle more, and the leſſe the better) but cut the graft thinne at the lower end of the cut, ſo that it may decline in one continued direct ſmoothneſſe, without dints, ridges, ſpaces or windings all along the ſlope, from on ſide of the Cyon to the other, otherwiſe it cannot joyne in all places to the ſtock. The graft being thus prepar'd, Lay the cut part of the graft upon the ſtraight ſide of the ſtock and meaſure juſt the length of the cut part or ſlope of the Graft, and with your knife take off ſo much of the bark of the ſtock, (but cut not away the wood of the ſtock) then lay the cut ſide of the graft upon the cut ſide of the ſtock, and let the ſhoulder of the graft reſt directly upon the top of the ſtock, ſo that the cut parts may joyne even and ſmooth all along the inſide of the barke of the graft, being placed upon the inſide of the bark of the ſtock, and ſo joyne them faſt together with ſome ſtrong Ruſhes or flags, and clay them on every ſide that noe Rain get in.

If the ſtock be very little the way of Grafting is the ſame, only excepted, that in this caſe there muſt ſome of the ſubſtance of the wood be taken away, that the graft in it's ſlope be not too big for the cut in the ſtock, in which operation ſo much there muſt be taken from the ſtock, that the inſide of the barke of the graft may anſwere the inſide of the bark of the ſtock, which being done, all things elſe are the former way performed. This is call'd whip-grafting, and

is

is apposed to the former, when no wood is cut from the stock: for shoulder-grafting 'tis required, that the stock exceed not in bignesse, for then the bark being taken from it there will not be a right application of the sap-channells of Cyon and stock required in the definition of grafting, the disbarked place in the stock necessarily being much greater then that in the graft. Yet if the stock be not 3 inches circumference it will doe very well. The one of these wayes is called shoulder-grafting, because the upper end of the downright cut is intended and made fit to leane as it were upon the shoulder of the stock: The other Whip-grafting, because the operator only makes his streight-down right cut and carryes not to indent it at all.

Some think this way fit only for great stocks: but I have grafted seedlings this way, so small that the Cyon was put in like a Wedg, and was very even to the stock on each side, neither stocks nor Cyons being neer an inch round: but if small plants are this way grafted, they must be tyed about after the former manner used in shoulder-grafting; the wound made by cleaving is very quickly made up, and cemented by the sap in grafting a young stock, whereas in old it is quite contrary.

The way of grafting in the cleft, has been of long use, and is generally known to all gardiners. The stock must be cleft in an even place, and the cleft so prepar'd with your knife, in the cleaving, that the sides be not ragg'd, both sides of the graft are to be cut down slopewise, and shoulders made or not made at pleasure; M. *Austin* well advises that the outside of the graft be bigger then the inside, unlesse the tree be big, but if it be so great as to pinch the graft much,

then

then to make the inner side thicker a very little, that so it may preserve the outside from being so pinch't, as to make the bark of the Cyon fit loose, and not receive the sap from the stock into the common channel, in such manner as is requisite for the begetting of a continuation between them.

There are other ways of grafting very excellent; as in a great tree, to prepare your Cyon as for the shoulder-graft, and then to take off so much of the bark, the head being before cut off as that the slope may just fit the disbarked place, as in some of the figures of Inoculation. Sometimes the Cyon being so prepar'd we raise up the bark, as in the other figures of Inoculation; but to cut it off fit, I count the best way, and have often practized with universall successe.

Ablactation is the same with grafting, saving that in that way the Cyon remaines on its own stock, and on the stock you graft together. For the stock you graft, being planted by the tree from which you have your Cyon, you disbarke and cut the Cyon, so that the inward part of its bark may answer the like disbarked place in the stock, so they being bound up together, and not seperated till you are sure they are surely incorporated, at which time the Cyon is cut from its own, and lives only by the other stock.

It is an ordinary imagination that by this way of Ablactation, Heterogeneous conjunctions may be made to prosper, but those that consider that the cause of the impossibility of dissimilar plants thriveing by any way of Insition, is not the difficulty of their first uniting, but the disability of the root and stock to nourish the head with convenient nourishment, will not easily admit such a fancy; Pears upon Apples, and Services; Apples upon Thorns, and the like

like plants will with ease take, and continue in good growth longer then such time as is required that the Cyon should depend upon the mother plant in Ablactation for the fastening of it till cementation be made; But after a perfect conjunction, and great shoots spring out, they (almost constantly notwithstanding the greatest care) will dye, which is an evident signe that this way can administer no help, it only providing that nourishment be not wanting to the first moneths, and not securing them from the danger of wanting for the future, fit and wholesome Nutriment for their maintenance and growth.

N. *What Plants take on different kinds.*

This is a generall rule for grafting, Inoculation, Ablactation, and conjunction by penetration, or any such way of propagation, that the Cyon or thing implanted be of like nature to the stock, to tell what neernesse in every kind is enough, is matter of great Art; 'Tis known that Plums will not grow upon Cherries, nor Peares upon Apples for many years, though for a while they may prosper.

I find that divers plants will take by enarching or Ablactation, that will not take by grafting; so Grapes, as the early red upon the great Fox-Grape; Apricots also and Peaches, which being secured upon their own stocks, will admit implantation unto another also, and take unto it, which by grafting I could never bring them to.

The strangest conjunctions that we observe to agree, are the Whitethorn with the Pear, Quinces with the Pear, the Pear with the Quinces, the Medlar with the Whitethorn, the Apricots with Plums that are

of

of full fap, and sometimes upon hard scurvy Plums, most use the White-Pear-Plums for that purpose; I find not but some other are as good (viz) the Primordian, Muscle, Violet. And it is true, that all roses cement and continue well upon bryers, as on the sweet-bryer, dogrose, I have Cherryes that grow upon Plum-stocks which is Sir *Hugh Plates* experiment from Mr. *Hill.* p.113. and Currans upon Gooseberries: what duration they may be of I expect to learn. I am not convinced by experience that Pears upon White-thorn are worse in their fruit but if so I shall preferre Apple-kernells before Crabs for a Nursery. I have tasted very excellent Katherine Pears without stone or hardnesse, that came from a Thorn-stock: nor were they smaller or harder (which Mr. *Taverner* asserts) then ordinary fruit upon the proper stock, however I advise that such as shall for want of Pear, use Thorn-stocks, that they graft very low, for otherwise the Thorn not growing proportionably to the graft, will cause the graft to decay, being never able to grow thereon, unto the bignefs usuall in Pear trees.

    There are almost infinite storyes of strange conjunctions which urge earnestly for credit, some of incisions made upon animall bodyes: The Lord of *Pieresch* had a present made him of a Plum-tree branch which bore blossomes and leaves, which sprang from a thorn that grew in the breast of a Shepheard, this Shepheard having got this Thorn by falling upon a plumtree. Raw silk has grown on the eye brow of a Lady, mentioned by *Borellus*, observ. 10 cent. 1. being drawn through the flesh to stitch up the lips of a wound there, and the growth was so considerable that it required

quired frequent cutting; and there was a Spaniard lately had a bramble that grew out of his belly. The improvement that from these and the like storyes, the Author in the cited place proposes, is, That with the blistering plaister the bodyes of divers beasts be excoriated and planted anew with silke, woole, or the like, where it may likely grow to the great advantage of the owners. When this has well succeeded, I shall propose another raritie from the first story (viz) That such who live about Glassenbury plant upon themselves some of that famed thorn that beares leaves on Christmas day; for if the button moulds, according to the story, made from the wood, kept their time of blowing upon the doublet, through the silke of the button, doubtlesse the plant grafted upon the flesh may grow through the very doublet too. Or in the mean time I shall wage on the successe of my improvment, asmuch as the observator shall doe on his.

N. 3. *Rules for Grafting.*

The time of grafting, possibly is any time of the winter; I have seen Apples grafted in *November*, & at *Christmas*, and yet thrive very well; but the best time is, that which immediately precedes the spring; if possible let the Cyons be gathered before the trees shoot their buds though some will grow now and then, notwithstanding they be sprouted, 'tis no matter though the stocks are budded; I have at Easter grafted above an hundred Apples and Pears without any fail.

The best way to keep grafts a long time, especially in pretty hot spring weather, is to wrap them all in wet mosse, or cover them with earth.

Lute is made with horse-dung & stiff clay well mix'd together; Mr. *Austin* advises, that in shoulder-grafting

ing, the Cyon may be put upon the West or South-side of the stock, because if so, those winds which are most dangerous cannot so soon break off the grafts as on the other sides.

If you would have a spreading tree, put in a long Cyon; if a straight tree, put on a short one, or let but one bud thrive.

Good bearing trees are made from Cyons of the like fruitfullnesse. Unbind grafts when they have shot great shoots, that the binding eat not into the tree, strengthen those that are weak with a stick tyed above and below the grafted place, like Splinters to a broken bone, till the cementation be made and confirmed.

If you would have store of any fruit quickly, cut off the head of an old stock, and graft thereon.

To Trees that bear great heads, and are of a fast and binding bark, such as Cherrie trees, some hard Apples, and other kinds of great fruit-bearing, and other plants, it is esteemed necessary by some to put in more grafts than one, least the sap finding not way enough, the tree receive a check and perish by the disappointment of the sap. However this reason may hold, certainly 'tis prudence to put in more Cyons than one in such trees, least that one failing, the stock likewise dye, being bark-bound and not able to put out a germen.

Cyons are best chosen from the fairest, strongest, not under-shoots or suckers, which will be long ere they bear fruit, which is contrary to the intention of grafting; the prime use of which I beleive rather to be the expediting, than the improvement of fruit.

N. 4. *Of Inoculation*

Inoculation is performed by takeing off that eye or little bud which containes the beginning of a bough provided for growth in the next spring, and planting it so upon another stock that the sap of the stock may without impediment or interrupt course passe unto the little eye (as I may call it) imperfect or inchoate bough, and serve it for Nutriment: For which operation the Bark must be cut either downright, with a cross cut on the top; the downright cut being about an inch long, and the cross cut onely big enough to serve for the easie lifting up the Bark: and then the sides being lifted up with a Knife or Quill, the Shield is to be put in, and the lips or sides of the Bark before lifted up, are to be bound down upon the shield: Or the cross cut may be in the middle, and then the shield is to be made picked at both ends (otherwise in the forementioned way, the lower end onely is made picked) and the four lips are to be lifted up for the letting in the shield. Others cut the Bark clean out in an oblong square, and cutting the shield exactly in the same dimensions and figure, apply it to the disbarked place in the Stock. Others cut their shield in the mentioned Figure, but take not off all the Bark answering the oblong square shield, but leave the lower part on the stock, under which they put the lower end of the Shield, and binde it down thereon. Other varieties there may be, and are used, some more of which are delineated in the annexed Figures: To take off the Bud clean from the Cyon, the best way is, to draw the lines of your shield through the Bark with your Knife, and to take off the rest of the Bark thereabouts, leaving onely the intended Shield thereon.

Having

The Exemplification of the Operations by the Figure.

a Denotes the ordinary cutting of the Bark for Inoculation.
b b The sides of the Bark, lifted up for the putting in of the Shield.
c The Shield taken off with the Bud, which lies under the Stalk of the Leaf cut off.
ln The Shield put into the Stock to be bound up.
d The Bark cut out in an oblong square, according to another usual way of Inoculation.
e The Shield cut out for the fitting the disbarked square.
m The same Shield put into the Stock.
f A variation of the fore-mentioned way, by cutting off the upper part of the oblique square, and binding the lower part down upon the Shield.
o The Shield so put in to be bound up.
e Another variation by slitting the Bark, that the Bud and Leaf may stand forth at *e*, and the Bark slit be bound down upon the Shield.
h A cross cut for Inoculation.
i The same cross cut lifted up in this Figure, somewhat too big.
k The Shield cut off to be put therein.
p The Shield put in.
g or *q* The cut of the Cyon or Stock for whip-grafting.
r r The cut of Cyon and Stock for Shoulder-grafting.
s The cut of the Cyons, and slit of the Stock for Grafting in the cleft.
t The Stock set for Ablactation or approach.
u The Cyon of the Branch for the same operation.
1 2 The Branch that is to be taken off by Circumposition.
3 The Branch that bears up the mould to the disbarked place.
4. The Branch of a Carnation to be laid.
5. The joynt where the slit begins.
6. The next joynt where the slit is propped open, with a peice of a Carnation Leaf put in.

Adde this at the 70th Page.

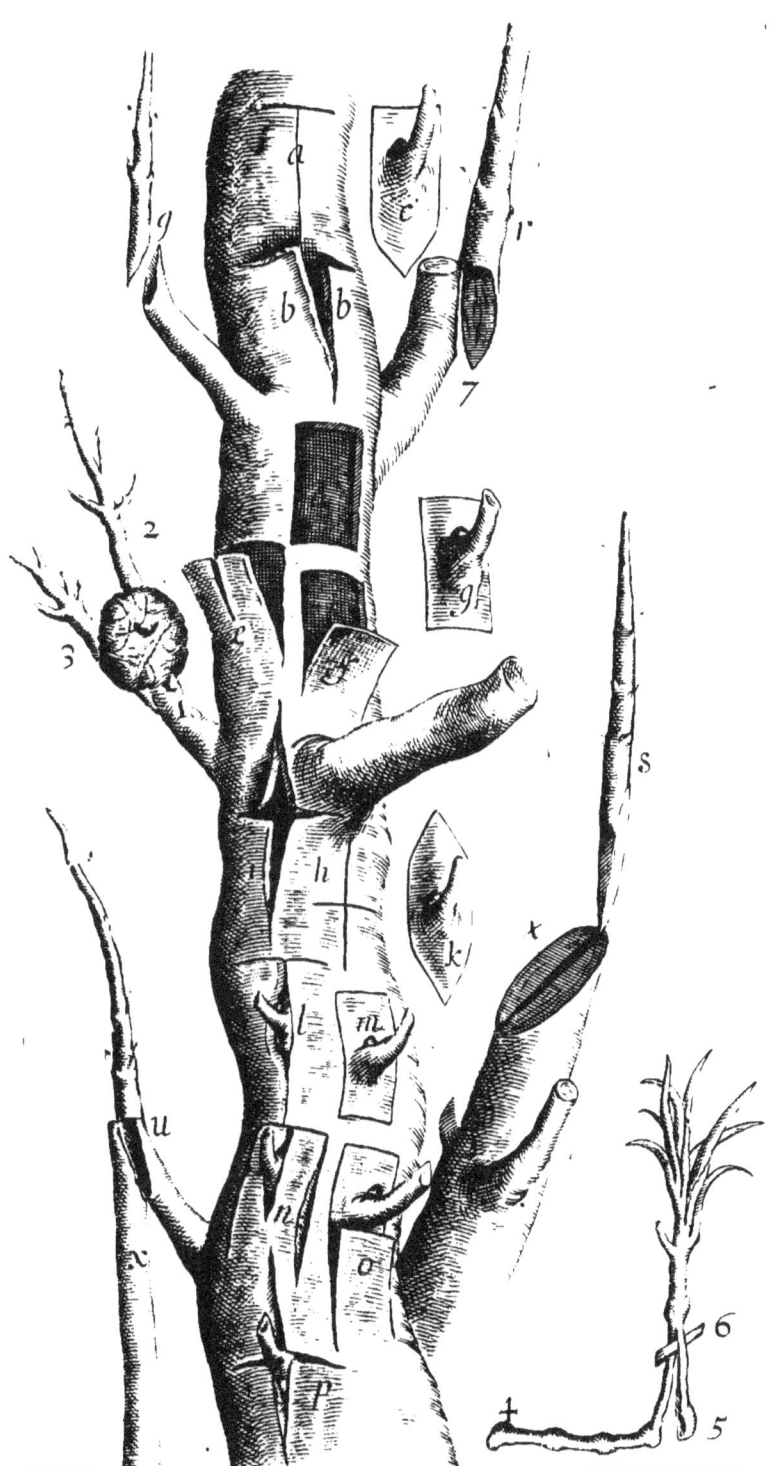

Having so far prepared your Bud before you take it off, remember to open the Bark of the Stock, for otherwise the shield will take hurt by the Air, which must be placed upon the Stock with all speed, and bound with something that may be of a yielding nature. The best way of taking off Buds, is with a Quill which is cut like a scoop, the one half, or two thirds, taken away for about an inch in length at the end: In taking the Bud off, be sure not to leave the Root behinde; for bindings, use any sorts of soft Rushes that will hold tying, long slipes of Linnen or Yarn.

I prefer such binding as need not be taken off till I expect the springing of the Bud, for there is much peril in premature loosing the bonds, yet 'tis necessary to unbinde whensoever the Stock swells about the place of Inoculation. The time of Inoculating is, from the first time you can get strong Buds that will come off after the frosts are gone in the Spring, till such time as that the Buds then implanted may be fast cemented before Frosts return in the Winter. You may Inoculate with the last years Buds, which are strong commonly, and fit to be put in at Easter.

Other Rules for Inoculation are, That the Cyon from whence you take the Bud be not weak, for then the shield will be so too, and likely bow or double in the putting in, which is a great reason why the double yellow Provence Rose is so hardly propagated by this means; other Roses, as the Rosa Muudi, Velvet, Marble, and Apples, Aprecotes, and the like, very easily, that the Bud be not sprung out much before it be taken off.

If you carry Buds far, expose them not to the
Sun,

Sun, but cut off the leaves, or some part of them, and wrap them up in wet Moss or fresh leaves, to keep them cool.

If the Bud take, in the *March* after cut off all that groweth above it, stripping away all the Buds that come forth elsewhere, or at the least all save one: some conceive one necessary for the drawing up the sap.

Choose strong Buds for Inoculation, and strong Cyons for grafting, and put them always on a smooth place of the stock.

Any thing may be propagated by Inoculation, unless the slenderness and weakness of the Shield hinder, that can be by grafting. Apples and Pears, though seldom Inoculated, certainly take. I have sometimes used to cut off the shield with a sharp knife flat, with part of the Wood thereto adjoyning, and put it in so; But this way, though many take, especially in Apples, yet the ordinary way seems better and more certain. Some take off Shields without a Quill, slipping them off with their fingers; but this is the ready way to leave the root of the Bud behinde on the Cyon, which being wanting, the other part of the Shield is unprofitable.

A pair of Compasses made flat at the ends, and sharp with edges, is an apt Instrument to cut away Bark for Inoculation, both for a true breadth and distance all at once; and so likewise with the same you may take off the bud truly to fit the same place again in the stock, Sir *H. P.* p. 113.

N. 3.

N. 5. Kirckers *Experiments concerning Infitions examined.*

Kircker, a Learned man, the *Pliny* of his time, after he had reproved the fallities in *Wicker*, *Alexius*, and *Porta*, who had afferted a change of colours and rare variety of flowers, by fteeping thofe roots in juices whofe colours were defired, feems to me as much to be blamed, in that he writes fo confidently of things which are as much like Paradoxes, and equally gain-faid by experience.

He fays, that he doubts not, but has from experience thefe effects; That a white Rofe, grafted upon a red, will bring that Rofe we call Rofa Mundi, or a Flower both red and white. This I have often prov'd falfe by mine own tryal: That a Gelfimine grafted on a Broom, will bring yellow flowers like thofe of the Broom; That I tryed, and could not make to grow, fo far it was from bearing any Flowers, *v. Kircher: ars Magn. p. 13. C. 6.* But that Jafmine upon Jafmine will grow and thrive, my own and others experience can atteft.

The fame Doctor, in another Book of his, *De Magnete*, where he has many good Experiments about that Stone, yet as to his φυτομαγνητισμος, either he is out, or there is greater difference betwixt the Countrey where he tryed his experiments, and *England*, then I can imagine; I have tryed Mulberies on Beech, Quinces, Apples, Pears, Elms, Poplars, and by grafting they would not take, yet he affirms they take eafily; and more, that Mulberies are by conjunction with white Poplars, made to be of a white kinde, and bear white Mulberies; That Pears

begin

being grafted on a Mulbery, being a red colour'd Pear, such I suppose we call the Bloody Pear, and that a Peach being Inoculated on it, it sends forth a bloody Peach, are his assertions, which conjunctions I see will not with us take, but if they would, I could promise my self no greater alteration of colour thereby, then I finde in the Flowers of Roses, which I have tryed in very many different sorts, and experienc'd to follow the Cyon without any participation of colour from the stock.

I having heard the same relation made of changing the colours of Tulips, by Artificial grafting the Bulbs of the white and red, and other colours, by proportionable indentments in each Bulb, tryed it this year in divers Roots, and made the Insitions, and put together the parts as artificially as I could, according to the rules here given; but the event is, that the Bulbs come not up at all, but die upon the operation.

Num. 6. *The maner of growing by Grafts.*

'Tis prov'd by experience, that there is every year a new coat of Wood made to every thriving Tree, by apposition of sap hardned into a thin Board (as I may call it) insomuch that I have known divers Woodmen, that would boldly assert the determinate number of years, that any Oke, or other Wood, has thrived in, by the number of those several distinct Rings of Wood that are to be counted from the middle or center of the Tree, to the outside of it, it being credited, and that I think with reason, that every one of these Rings arose from the apposed and hardned sap of every several year.

Now in grafting upon a faim stock, it comes to pass

pass, that the sap of the stock is apposed to the body of the Cyon, and so incloseth the Cyon with the last coat of the whole Tree, that there is, as it were, one and the same past of new Wood, that doth closely encompass the whole, both Stock and Cyon, which when harden'd, grows to be strong, and of the same use that splinters are to a broken Bone; and Gardners wisely provide for the strengthning of the compagination of the Cyon and Stock, until this sap be incrusted to a hardness; when the first year of their grafting, they do not onely binde up the Cyon to the Stock, but use splinters of old Wood, that neither the winde, or other accidents, may dislocate what with Art was joyned together. This first, for the maner of conjunction and fastning of the Woods: Nor do I make any difference between Grafting and Inoculation, because I am perswaded, that as there is in every Seed an actual Plant, so there is in every Bud an actual Bough, and that a Cyon and a Bud differ but as a greater and lesser branch.

But how the sap of the Stock; suppose White Thorn can serve to make the Wood, Bark, Leaves, and Fruit of its Cyon, suppose a Pear, is a difficult question: For grant there be an elective attraction of sap from the earth; yet how shall a white Thorn choose that which is fit for a Pear? My thoughts are, that for those who maintain election of similar parts, it were best to suppose a great likeness in all Grafts and Stocks, as to their inward nature and parts, though not outward figuration; and there being this likeness in the substance, it will not be hard to conclude, that the Cyon, by altering the position of the same substantial parts, may make to the sight, smell, touch, taste, a thing of another fashion.

For

For those qualities that affect the senses vary often in one and the same thing: The Apple in the beginning that is without smell, of sowr taste, green colour, hard to the touch, shall in a little space be fragrant to the Nose, sweet to the Palat, of a golden or ruddy colour, and soft to the feeling: And in a thousand instances 'tis found, that several positures of the same parts, shall produce several opposite colours, and other sensible appearances in the same thing: There is no inherent colour, either in the infusion of Galls or Vitriol (though limpid they are not) so dark or deep as to come near the blackness of Inck, which notwithstanding, being mixt, they produce it. Two other infusions of like colour, would not upon mixture arise to such an effect, because not able to dispose each others particles into such positures. Spirit of Vitriol, though without colour, disposes the parts of this Inck so as to destroy the blackness; Oyl of Tartar restores both position of parts and pristine colour; and that it arises from different positures, may be argued, because there is a visible motion, striving, and local mutation in them, before these last effects are produced; and 'tis plain, that when the Inck, by reason of the spirit of Vitriol, disappeared, yet all the parts were there, for else it will not be imaginable how a limpid Liquor, as Oyl of Tartar, should reduce the colours which it does not by it self generate, as it is plain, because restoring Letters written with Inck, and taken off with Spirit of Vitriol, it makes no blackness on the Paper, save onely upon the Lines of the Letters: These two limpid Liquors likewise, being put together, turn into a good consistence and milky colour.

But

But he that desires more instances of this kinde and matter, that, according to this Doctrine, may much help the Theory of Colours, and particularly the force both of Sulphurous and Volatile, as likewise of Alkalizat and acid salts, and in what particulars Colours likely depend not in their causation from any salt at all, may beg his information from that Noble person (in order to whose command, for all his intimations to me are such I am now writing) who has some while since honored me with the sight of his Papers concerning this subject, containing many excellent Experiments made by his Honor for the elucidation of this Doctrine; or otherwise, for the present, may see very good instances hereof in Dr. *Willis* his Treatise *De Ferment, cap.* 11. And truly, if Tastes, Colours, Smells, were not easily alterable, it would not be that we should from the seed of the same Plant attain to such change and variety of Flowers and Fruits as are mentioned above, nor of Flowers from the same off-set.

But if there be supposed in the world, and all several Bodies, but one Element or material principle, from which by Natures undeseryed Wisdom, in appointing it into several motions and changes of scituation, and giving different Measures and Figurations to its smallest Particles, there arise all the varieties in the world, then there will be no difficulty how the same sort of matter should give substance both to the Stock and Graff, though Plants of different nature, and bearing different Boughs, Leaves, Fruits, Seeds, each from other; for if from any matter, any thing may be made without difference, then particularly the wildest Stock may afford Elements fit to nourish the Boughs of any Plant, of

how

how gentle and noble nature soever.

But lastly, If all these Considerations be too troublesome, I can help a lazy Naturalist to an admirable expedient for the resolving this appearance; let him be content to believe, that when the Sap, gather'd in the Root, comes to the place of conjuncture, it is there forc'd to undergoe a total corruption and lapse into the Bed of its first matter, from whence, by a new generation, there arises a new sap, begot in the Tree by a specifick faculty, which in a Pear graff may be call'd a Pear-sap-making-power, and so in all the rest: And for the commendation of this last way of Resolution, I must express this its excellency, that it is equally applyable to all things in the world, each thing being made (and the cause as easily believed) by some such thing-making power.

Or it might not be amiss to entitle *Diva Colchodea*, the grand-general form-making-intelligence, to the production of all these effects, and in Romantick guise, to place her, as it were, in a non-erring chair, sitting in the very place of conjuncture of Cyon and Stock, and working by ways and arts belonging to her own Trade (and therefore, as her proper mysteries, not to be revealed) to the forming in most occult and admirable maner of the appearing effect.

CHAP.

CHAP. VI.

Of the ways for, and Seasons of setting Plants.

ALl Trees and Shrubs of Woody substance, that have Bodies able to endure the cold, are best set before the Winter, assoon as the Leaves begin to fall: A Quickset of this season, will far outgrow the like planted in the Spring. Artichocks and Asparagus Roots do exceeding well, being planted at this Season, if set in a rich warm mould, and well defended in the ensuing Winter from the violence of the frosts: Artichokes are with us set above an Ell distance, and thereby in the Winter, a Trench being made between the rows, the Mould is cast up on ridges for the defence of the Roots; and in the Summer, Cole-flowers, or other Garden-stuff is set in the distances. For Herbs and choice Plants, especially those that are set without Roots, it is most fit and usual that they be set in the Spring, as Hysope, Time, Savory, Marjerome, Wall-flowers, Pincks, Gillyflowers and Carnations, with this Caution; That by how much more tender each Plant is, in regard of cold, the later it requires to be set, and in the warmer place.

For all bulbous and tuberous rooted Plants, it is accounted the best way for their preservation and improvement, that they be taken up every year out of the ground, and kept some time out of the ground. The Universal and Catholick order of all Bulbous Plants,

Plants, says *Laurembergius*, is, that about St. *James* tyde they be taken out of the ground, and put in a place cold and dry, of a free air, not in the Sun, nor covered with Sand or Earth, or accessible to Mice; let them abide so a Moneth, or thereabouts, then set them again, when they are taken up, cut off the Fibres that grow from under the head: nor need any thus take them up every year (unless it be for the transplantation of the off-sets) by which forbearance, the stock of Tulips is very much increased. *Ferrarius* more particularly forbids the abiding of Anemones in the Earth all the Summer, as being found prejudicial to them by his experience. But Fritellaries, and Peonies, and the Crown Imperial, he will not have removed from their Beds, unless into a Cellar, in a pot of Earth.

Nor are all taken at the same time, as he seems to intimate; for Narcisses and Crocusses are commonly taken up first, generally when the flower is gone, the leaf withered, and the Bulb full, it is the best season to take them up: some keep them out of the ground longer, as till Christmass, or after; as this year, being in *London*, my best Tulips, Anemones and Ranunculus's, were in the House till the beginning of *February*, and yet did well enough: But commonly we re-plant them about Michaelmass, or thereabouts: some great Florists keep them out of the ground no longer than till they grow dry; some replant them in *June*, some in *July* or *August*; some take not up their Ranunculus Roots at all. Those Gardiners, whose Beds are apt to be over-flowed or soaked with cold water in the Winter, the later they set, I believe their Bulbous and Tuberous Roots will prove the better.

The

The ordinary time to plant Anemones, says Mr. *Parkinson*, is most commonly in *August*, which will bear Flowers, some peradventure before Winter, but usually in *February*, *March*, and *April*, few or none abiding until *May*: But if you will keep some Roots out of the ground un-planted till *Febr. March & April*, and plant some at one time, and some at another, you shall have them bear Flowers according to their planting; those that are planted in *Febr.* will flower about the middle or end of *May*, and so the rest accordingly, & thus you have the pleasure of these Plants out of their seasons, which is not permitted to be enjoyed by any other that I know, Nature not being so prone to be furthered by Art in other things, as in this, yet regard is to be had, that in keeping your Anemones out of the ground for this purpose, you neither keep them too dry nor too moist, for sprouting or rotting, and in planting them, you set them not in too open a Sunny place, but where they may be somewhat shadowed.

N. 2. *Of the setting of Woods Fruit-Trees, and Plants uncultivated.*

Concerning Plants that are ordinarily set abroad, and are not cultivated in Gardens or Orchards, few observations can be made that are no: very vulgar; 'tis greatly his interest that mindes the thriving of his Trees, that they be set that the Roots may run just under the Turf, in the surface of the Earth, the higher the better, if they are kept moist at the root with wet straw, or the like, and defended from injuries the first year. I have seen soom plants so buried in a depth of thick clay or gravel, that they could not shoot for many years a sprig of a Span long, whereas others set orderly in the same place did well upon a Stone Wall, that is washed with rain Water, as in that hole, when once the Root is come to the sides thereof. This I speak generally and not of such particular Trees as delight in a singular Mixtera of Earth.

And for Orchards, it is a very necessary requisite, that the Roots of Fruit-trees stand above the Gravel, Clay, or Rock, if any such be, provision for which I have known made two ways, the usual and most common is, to plant with such Standards which have no down-right Roots, which may be gotten in any well ordered Nurseries; for in such, the Seedling Plants are taken up the second year, and the down-right Roots being cut off short, they are set in beds for grafting, and by this means shoot their Root rather in compass, then directly downwards. The second way is a more unusual experiment (*viz.*) To set the Fruit-Tree on the top of the ground, without any hole dig'd, & to lay a load of such dirt as is found in streets to the root, upon the Turf; yet so, that the rain may abide, and not by reason of the banck, run from the root of the new set fruit-tree.

For Wall-Trees, it is convenient the Roots be set at such distance from the foundation of the Walls, that they may have room in the Earth for their roots; a foot is a convenient space generally; for then the heads will without difficulty be drawn to the Wall, and the Roots not be prejudiced.

Those Wall-Fruits that are set abroad, as Vines, &c. being kept short in their Branches, and not suffered

fered to climb, become good bearers, especially if they are set near the reflection of Gravel-Walks, or upon other Ground kept bare from Weeds.

For the planting of Woods in general, for increase of under-Wood, Mr. *Blith*'s way is generally approved, to cast up double Ditches, and plant any sorts of Wood in the form of a Quick-set: Some sow seed on the Bancks in orderly rows, and set likewise on the top, as well as both sides of the Bank. The time is, assoon as the Leaf is fallen, in any Weather or Season. The Plants in a more sound ground, are Ash, Oak, Elm, Sycamore, Maples, Crabs, Thorns; in a more moist Ground, as a drained Bog, Poplar, Willow, Sallow, Osier, which grow by Truncheons. In which watery soils, the way of raising Ditches is most necessary: For neither Willow, Sallow, Osier, nor any other Plant, will grow in a Bog, without soundness of ground: What Plants grow by cuttings, what by laying for the more ready thickning of Woods, may be seen above in the proper Chapter.

There is a story freely defended and frequently, both in discourses Printed and spoken, that the chips of Elm, being sowed, will grow; but that is somewhat like *Kirther*'s experiments, before-mentioned in the Chapter of cuttings, and not a whit more true; otherwise, to sow those Chips would be a good profitable and frugal way for thickning Woods. The cause of the Countrey mans mistake (for I suppose not that this error arose from Philosophers) I imagine to be this: At the felling of great Elms many chips must needs be scattered, and flie round about the Tree, and be covered in Grass thereabouts; now the next year, after the fall, there arise generally great numbers of Suckers from the roots of the old

Tree,

Tree, which roots, must emit all the sap they gather up into these Suckers, the great Trunck being removed. And these Suckers are easily mistaken to arise from the chips, because they always come upon the felling of Elms where chips are found, and grow at such distance as chips are ordinarily scattered.

N. 3. *Whether any Vegetables may be set so as to grow in the Air.*

There is a question now-adays frequently proposed, Whether there be more Soils then the ordinary Turf or surface of the Earth, tempered with some water, soyl being meant for the ground, in which things may be set to grow. I need not speak much upon it, as to Water, which by Experiments related in the Chapter concerning Propagation by cuttings, appears to have a property to elicite Roots, and make them where they were not, and nourish the Plants by them after they were made; to which, I must adde this circumstance, not before mentioned, that Periwinckle, and divers others, continued their growth by this nourishment alone, from year to year, not dying in the Winter. How long they might have continued, I can't assert, for being absent this Winter, and no fires being kept near, the water in the Glasses, was so raryfied by the Frost, that the sides could not contain it, but were forced asunder thereby, and so the Plants perished; whereas otherwise, they being set in a Room over my Laboratory, I question not, had many of them continued till now.

Some put forward, that the Air might have the faculty of nourishing Vegetables ascribed to it: And no wonder, when *Paracelsus* makes it a
sufficient

sufficient nourishment for men, and brings instances for the proof of his assertion. But I finde, That Onions, Tulips, and all Bulbous Roots, though they shoot out a green leaf, yet do very much lessen in their weight, and it appears, that this growth is but the motion of the same parts, or some few of them, to settle and gather in another place, and another order or scituation in relation to each other; for the Onion particularly hath the thicker coverings of the Bulb very much stretched out, and each covering, as it increaseth in length and breadth, by rising into a leaf, so the thickness considerable while it covered the Bulb onely, decreaseth proportionably and is molden into a thinner, and more largely extended Vestment.

I have hung up divers Sedums, Orpines, Tithymalls, and other such Plants, which I imagined most likely to grow by the Air onely, and to encrease and be augmented thereby, and found, that by all my endeavors, though the Plant grew well, yet they always lost weight, and never got the fourth part of a grain.

Aloes likewise, though being hang'd up in the air with a cloath dipped in Sallat Oyl, it sends forth for many years new leaves, yet it always grows less and less in weight, till at last the oldest leaves falling off, and new coming up, it grows to nothing.

CHAP. VII.

Of the means for the improvement and best culture of Corn, Grass, and other Vegetables belonging to Husbandry; and of the ways for removing the several annoyances that usually hinder such advantage.

Num. 1, *Of the Annoyances to Land, and the Impediments that usually distemper it, to the disadvantage of the Husbandman.*

THe Impediments that with us hinder the Husbandmen from making the greatest advantage of their ground, are either the distempers of the ground it self, or some evil accidents that occasionally happen thereto, or to the vegetables growing thereon. The distempers are generally caused, either by the abounding of water, above all other principles, which causes coldness, and a Dropsical disposition in the Earth; or by the abounding of a dry Earth or Mineral, and the want of moisture and saltness, and that Spirit which should cause

that

that motion in the insensible particles of the Earth, which is proper for the exciting the Seeds of all things, and so stirring the ground, that the several particles may be at liberty to enter the Bodies of Vegetables fit for them; the accidents come by blasting Winds, rapacious Fowls, Vermine, and Weeds, Fearn, Heath, Broom, and other improfitable Vegetables; of these, and the usual remedies against them, somewhat, and the best that at the present occurs, I shall speak in this Chapter.

N. 2. *Of the remedies proper to cure the excessive coldness and moisture in Lands, and the ways of Improvement thereby, in Grounds subject to these distempers, by dreining, Pigeons and Poultry dung, Urine, Soot, Ashes, Horse and Sheep dung: Of Ground cold and dry, and how these Soyls may be applyable thereto.*

Bogginess and obstruction of Springs more or less, is generally the cause of the chill or coldness that lies upon Lands, and breeds the Rush and other incommodities, and therefore the foundation of the cure, and improvement thereby, must be to remove this internal cause, by laying the ground dry, and dreining the Bog: In the relation of which operation, and many more of this Chapter, I shall ease my self, by giving you Mr. *Blith*'s observations and directions thereabouts, who was both a Practicer himself, and questionless a very faithful and true Reporter of his experience.

In cold, rushy land, says he, the moisture, or cold hungry water, is found between the first & second swarth of the Land; and then oft-times you come immediately

ately unto a little Gravel, or Stonyness, in which this water is, and sometimes below this, in an hungry Gravel, and many times this Gravel or Stonyness lieth lower: But in boggy Land it usually lieth deeper then in rushy; but to the bottom, where the spewing Spring lyeth, you must goe, and one spades depth, or graft beneath, how deep soever it be, if you will drain the Land to purpose.

And for the matter or Bog-maker, that is most easily discover'd, for sometimes it lieth within two foot of the top of the ground, and sometimes, and very usually within three or four foot, yet some lie far deeper, six, eight, or nine foot, and all these are feazable to be wrought, and the Bog to be disovered; but until thou come past the black Earth, or Turf, which usually is two or three foot thick, unto another sort of earth, and sometimes unto old Wood, and Trees, (I mean the proportion and form thereof, but the nature is turned as soft and tender as the Earth it self) which have lain there no man knows how long; and then to a white Earth many times, like Lime, which the Tanner & White-Tawer takes out of their Lime-pits, and then to a Gravel, or Sand where the water lieth, and then one Spades depth clearly under this, which is indeed nothing else but a Spring, that would fain burst forth at some certain place, which if it did clearly break out, and ran quick and lively, as other Springs doe, your Bog would die, but being held down by the power and weight of the Earth, that opposeth the Spring, which boils and works up into the earth, as it were, blows it up, and filleth the earth with winde, as I may call it, and makes it swell and rise like a Puff-Ball, as seldom or never you shall finde any Bog, but it lieth higher, and rising from the adjacent Land to it, so that I believe

lieve, could you possibly light of the very place where the Spring naturally lieth, you need but open that very place to your Quick-Spring, and give it a clear vent, and certainly your Bog would decay; by reason whereof, it hath so corrupted and swoln the earth, as a Dropsie doth Mans Body; for if you observe the mould, it is very light and hollow, and three foot square thereof, is not above the weight of one sollid foot of natural Earth, Clay, or Land, whereby I conceive, that how much soever this mould is forced from the natural weight or hardness of solid Earth or Clay, so much it is corrupted, swoln, or increased and blown up, and so much it must be taken down, or let forth, before ever it be reduced; I therefore prescribe this direction: Go to the bottom of the Bog, and there make a Trench in the sound ground, or else in some old Ditch, so low as you verily conceive your self assuredly under the level of the Spring or spewing water, and then carry up your Trench into your Bog straight through the middle of it, one foot under that Spring or spewing water, upon your level, unless it rise higher; as many times the Water or Spring riseth, as the Land riseth, and sometimes lieth very level unto the head of the Bog, unto which you must carry your Drain, or within two or three yards of the very head of it, and then strike another Trench overthwart the very head both ways, from that middle Trench, as far as your Bog goeth, all along to the very end of it, still continuing one foot at least under the same, and possibly this may work a strange change in the ground of it self, without any more Trenching.

Or thus you may work it somewhat a more certain way, but more chargeable, (*viz.*) After you have

brought

brought a Trench to the bottom of the Bog, then cut a good substantial Trench about the Bog, I mean, according to the form of your Bog, whether round, square, or long, or three or four yards within your Boggy ground; for so far, I do verily believe, it will drain that which you leave without your Trench, at the depth aforesaid, that is underneath the Spring-water round; And when you have so done, make one work or two just overthwart it, upwards and downwards, all under the matter of the Bog, as is aforesaid, and in one years patience, through Gods blessing, expect your desired Issue: And if it be in such a place as will occasion great danger to your Cattle; then having wrought your works and drains as aforesaid, all upon straight lines (by all means prevent as many Angles, Crooks and Turnings, as is possible, for those will occasion but stoppages of the Water, and filling up of Trenches, and loss of ground, and much more trouble then otherwise.) Then you must take good green Faggots, Willow, Alder, Elm, or Thorn, and lay in the bottom of your Works, then take your Turf you took up in the top of your Trench, and plant them thereupon with the Soard downward, and then fill up your works level again, until you come to the bottom or neither end of your work, where your Trench is so shallow, that it will not endanger your Cattle; or rather take great pibble Stones, or Flint Stones, and so fill up the bottom of your Trench, about fifteen Inches high, and take your Turf and Plant it as aforesaid, being cut very fit for your Trench, that it may ly close as it is laid down; and then having covered it all over with Earth, and made it even as the other ground, wait and expect a wonderful effect, through the blessing

of

of God; but if you may, without eminent danger, leave your works open, that is most certain of all.

Other and second remedies for all cold Land, are Pigeons Dung, Dung of Poultry, which abound in heat and volatile salt; these are onely sowed by the hand, for fear of burning the Corn in the chitting of the Grain: I have observed, where these Dungs have been over plentifully laid, that the place bare no Corn at all, when as in other places, where it was moderately strawed, the Crop was exceeding great. The same effect there is in Urine and Soot, from the same principles, (*viz.*) much eager spirit and volatile salt, and therefore the same caution is to be had in their use. I have seen half the Trees in a Codling-hedge killed, by watering them over-much with Chamber-lye. Horse-dung, if not rotten, lying thick, will doe the same, but rather by an actual heat which it creates by its fermentation, than by the power of single principles, as in the former instances, but the excess of it is harmful, being laid in such quantities as it may heat, and certainly burns the root of any ordinary Vegetables that grow near it. Sheep-dung, Hog-dung likewise, and all Soyl and Litters of Cattle, by reason of their Dung, Urine, and heat of their Bodies, lying thereon, have a warmth in them, and are fit for cold Lands on that account; and by reason of their moisture, for dry Lands also; for it is to be observed, that many Grounds are dry and cold too, in all parts of the North and North-west, as *England* lies, and in *England* many of our Wood-lands especially; and so all hot and moist soils are most proper for them: Burning and beaking is in many places very successfully used to this effect; The actual fire heating the ground,

ground, and the ashes of Fern, Brake, Heath, &c. of like nature, yielding a salt, very profitable for, and expedient to joyn with the other principles in the ground, to cause a fermentation and fruitfulness.

'Tis a general rule, that there is nothing in animal Bodies, but will turn to excellent Manure: Their Horns, Bones, Hair, Flesh, both of Beasts, Fish, and Fowl, are very rich; and those that know the vertue of them, buy at Cities for the purpose, rags which are made of Wool, Sheep-trotters, stincking Fish, or other Offal of Animals, which must either be mixed with other Dung, or not laid over thick.

But it is to be observed, That where moisture is rather required then heat, there floating by Landfloods, the dirt and mud of Ponds and High-ways is most proper: where warmth and heat, is a greater need; there soyl that is made by a mixture of the Offal of Animals, will be more to the purpose and advantage of the Husbandman.

Lastly, 'Tis probable that any thing that has active parts in it, if it be not just of the nature of the ground, will raise improvement: Heterogeneous things, upon their meeting, ordinarily causing that stir, which is thought, by most Naturalists now, to have great influence upon Vegetation.

N. 3. *The ways of Improvement of dry, light, sandy, gravelly, flinty Lands, by floating, Marl, Chalk, Lime.*

Dryness is generally a great cause of barrenness, and is an usual annoyance in Sandy and gravelly grounds,

grounds, more especially, in regard that they retein not the rain-water so well as clay, or Land of a mixt soil: The proper remedy for this defect, is artificial watering, which tempers the ground most properly for the improvement of the growth of the most useful Plants, Grain and Grass: For first, Water in its own nature and property is a soil, and has an exceeding agreeableness with the Bodies of most Vegetables, as appears by the experiments of their growth in water onely. And secondly, There is a very considerable accrewment to dry, sandy, and gravelly Earth, by the fatty soyl and wash that is carryed both in Land-floods, and other Water, that having passed through Cities, Roads, or other places of like nature, are drawn over the ground, for the salt, and other the mixt earth, that was carryed in the Flood, being apt to reside to the bottom, is left generally behinde upon the Land; and the salt diluted in the Water, easily enters the Turf, and carries with it other Particles thither, where, by the heat of the Sun, (they being in conjunction with the Sand, Gravel, or other Bodies Heterogeneous, and unlike to themselves) they cause by their mutual fermentation, as is supposed, or some other way, that temper of ground which is most fit for the growth of all Grain, Grasses, and other Vegetables of general use.

For drawing the water over Land, the use is, that by the eye or level which is easily made to help the eye: First, Discovery be made where the water may be conveighed over the most Land: Then Mr. *Blith* advises, to cut out the Master Trench or Water-course, to such a bigness, as may contain all the Land-flood, or at least, be able to bring it within the Land intended for this improvement: When the

water

water is brought thither, carry it along in a foot broad Trench, or lesser, all along the level: If the level be too dead, the lesser stream will follow, so that a convenient descent must be minded, to give the water a fair passage. If there be discovered in this lesser Trench, any mistake or failing, it may with ease be amended, by going higher to, or lower from the level, and the first Trench be stopt up again; for this Trench need be no deeper then the thickness of the upper Turf: This done, the Water-course must be cut out, which must be large enough to contain the whole Water which is intended for the enrichment of the Land, which largeness ought to consist in breadth, and not in deepness, for a shallow Trench, about a foot deep, is best for this work: When the Trench is brought near to the end of the Land, it is to be drawn narrower and narrower.

Further directions the Author gives the Improver, in these words,

"As soon, says he, as thou hast brought the Water upon the Land, and turned it over, or upon it, be sure thou take it off as speedily as possibly, and so fail not to cut thy work; so as unlesse thy Land be very sound, and thy Land-flood very rich, thou must take it off the sooner by a deep draining Trench. Therefore I prescribe no certain breadth, betwixt floating and draining Trenches; but if the Land be sounder and dryer, or lieth more descending, thou must let it run the broader; and as the Land is moist, sad, rushy, or level, let it run the lesser breadth or compass; and for the draining Trench, it must be made so deep, that it goe to the bottom of the cold, spewing, moist Water, that feeds the Flag and Rush; for the wideness of it, use thine own liberty, ty, but be sure to make it so wide, as thou maist goe to the bottom of it, which must be so low as any moisture lieth, which moisture usually lieth under the upper and second swarth of the Earth, in some Gravel or Sand, or else, where some greater Stones are mixed with Clay, under which thou must goe half one Spades graft deep, at the least: Yea, suppose the corruption that feeds and nourisheth the Rush or Flag, should be a yard, or four foot deep, to the bottom of it, thou must goe, if ever thou wilt drain it to purpose, or make the utmost advantage of either floating or draining, without which, thy Water cannot have its kindely operation: The truth is, otherwise the benefit might happen to be no greater then the Patients, who incurr'd a Dropsie in his cure from a Fever; whereas by this means there is a double benefit; the first whereof comes by the commodity of watering; the second, by the dreining Trenches necessarily annexed thereunto: And whereas the aforesaid Author commends watering or floating as an help to boggy, rushy, quagmiry Land, I suppose no benefit, but hurt would arise thereby to such Lands, if these dreining Trenches did not open the passages of the obstructed Springs original causes of the Fog or Rushiness, as well as let out the Water newly introduced by the floating.

The time of the operation for this improvement, must be when the Grass is all off the ground, for else the soil will stain it that comes along with the Flood: Often watering is good, but to keep it long in a place, breeds the Rush. By this very Husbandry, Mr. *Blith* brings precedents of improvement of Land, from Eighteen pence, to Thirty shillings

an

an Acre; and Mr. *Plat*, from One shilling to Five pounds.

Another remedy for dry and light ground, such as abound in Sand and Gravel, is Marl, an Earth most commonly slippery, or greasie to the touch, sometimes blew, sometimes grey, otherwhiles yellow, now and then red, always fryable, so that it will slack after a shower, and not grow afterwards hard or crusty, as Clay doth, but easily resolves to a dust or powder: It saddens Land naturally, and so will turn Rye Land as to make it fit for Wheat, Barly and Pease, and therefore must not be used twice or thrice together, without some other more rarifying compost to intervene, such as ordinary Dung is; if you lay it down from Tillage, 'tis requisite that all Marled Land be first well dunged.

Chalk also I have seen used with very good success in *Hampshire*, upon the Downs there, which are of so dry a nature, that it is grown Proverbial there, that their Ground requires a shower every day in the Week, and on the Sunday two; and Mr. *Blith* affirms, that in *Hertfordshire*, by Chalk, the Improvement is made on Barren, Gravelly, and Flinty Lands.

Mr. *Blith* reports thus of Lime, that it is a suitabler Soyl for light sandy Earth, then for a warm Gravel; 'tis improper for a wet and cold Gravel, but for a cold hungry Clay worst of all; for, says he, Lime being once slacked and melted, is of a cold nature, and will sadden exceedingly, contrary to its nature, in the Stone, for it turns light Land into such a capacity, that it will bear exceeding good Lammas Wheat, or mixed Corn: About twelve or fourteen Quarter of Lime serves an Acre, it may as well be

over,

over, as under-limed; after Liming, till not long, but return to Pasture.

Num. 4. *Remedies for accidental annoyances and hindrances of Improvement, particularly the ways to destroy Fern, Heath, Ant-hills, Moss, Rushes, Rest-harrow, Broom, or any such Weed or Shrubs that infest the ground: Whether liming of Corn prevents blasting, the effects of that and Brine in Improvement: Concerning Moles, and the ways to destroy them or drown them; a way of Antipathy, as to this effect, in Animals and Vegetables to the Bodies of their own kinde, when they are in the way of corruption: Mr. Blith's way of preserving Corn from Crows, Rooks, &c.*

When any Land runs to Fearn, Heath, or Ant-hills, Mossiness, Rushes, coldness, or any other Weeds or Shrubs, as Goss, Broom, Furz, &c. The most proper and improving remedy, is, to plow it three or four year, and then lay it down in good heart. In which operation, care must be had to plow up the Weeds clean, and burn the Roots of them in heaps, which warms the ground, and to give it convenient dunging every year, for so the greater shall the improvement be. This Land must be cast into Furlongs, that the Furrows may convey the Water one to another into a general Trench, that it lie not upon the Land.

If the Land be cold and moist, lay it the higher on ridges; if hot and dry, sandy, or the like, let it lie flat, that it may better retain the Rain water.

Be sure you Plow up the Rushes, Brakes, or other annoying Weeds, and for fail let some body, with a Spade, follow the Plough, to root up such as are left after the Culter and Plow-share.

Harrow this new broken ground with weighty, sharp, and long tined Harrows, such as 'tis a Teems work to draw, that uneven places may be torn up, and good store of mould raised. Cover your Seed with two or three sorts of Harrows, each Harrow having tines thicker then the other: some put weights upon the Harrows in the first, and a Thorn under them in the last operation.

After four years Tilth, lay down your Land, and that upon a Crop of Wheat or Rye, not on a Summer Corn, for so the Soard will come the sooner, especially if the Crop be sowed thin, and as early as may be: If you will double or treble the Improvement, the Husbandry of sowing Clover-grass, spoken of in the first Chapter, will here come in most properly. This last Plowing, regard that the Ground be laid down smooth, yet on ridges if the Land be cold, and unless the Land be of exceeding strength, fail not to manure it, by dung, or otherwise, this last season of plowing.

Mr. *Blith* reports, and Mr. *Hartlip* likewise, That the natural helps to preserve Corn from blasting, is the steeping of it in thick fat water, or Lime water, Urine or Brine, or the mixing of Lime or Ashes, with Corn well wet and moist, that so it may cloath it self with the finest of the Lime or Ashes, &c. so as it may fall cloathed all over to the Earth, and so be covered therewith: But I believe he was mistaken in the applying of the Medicine to the prevention of the right and proper disease: I have heard such who
practiced

practiced these Medicines, affirm, that they have generally, and with reasonable good success, used those remedies to prevent smootiness; but the very last year it was observed, that where those means were used, the blast did as much harm, as on the adjoyning Lands, where there were no such Applications made to the Seed. And blasting being the perishing of the tender Kernel, by reason of a Wind (which from the effect is sometimes called a red Wind) that too sharply, and it may be with some Venome breathes on it at its first beginning; I see no reason that such infusions or applications should be any defence, for it comes from an outward violence, and therefore it is most usually seen, that not half a Tree onely, but half a bough shall be blasted, while the other half of the same, that grows by one and the same nourishment, remains free, sound, and well coloured.

There is a procedure mentioned among Mr. *Speeds* notes, for Liming Corn that carries a good probability of advantage with it. First, The Grain was steeped in strong Brine of Salt, that would bear an Egge twenty four hours, and then being laid S.S.S. with Lime that is there, was laid a layer of Corn first, and then a layer of Lime, and then again a layer of Corn, &c. the Lime cleaved to the Wheat, and was sowed on Ground not worth Two shillings an Acre; the effect was, That it bare as good a crop of Wheat as ever was seen in *England*, and afterward three Crops a year of Clover, exceeding good, one whereof was equal in value to a Crop of Wheat: This being matter of Fact, I believe it, as to improvement by fertility, because the Brine works very considerably in small proportion, and Lime in this juncture may do well, both to fertility, and defence of the Grain against Grubs, and Insects, and Worms, that abide in the Earth; but surely as to blasting, and Crows and Birds that spoil the Corn in the Ear, it has no influence.

Moles by watering are drowned, or driven up to so narrow a compass, that they may be easily taken; I have known them to have been forc'd to leave their holes to run upon the Turf, to save their lives from the Water-flood. Mr. *Blith* relates, That one Spring, about *March*, one Mole-catcher and his Boy, in about ten days time, in a ground of ninety Acres, being just laid down from Tillage, took about three Bushels, old and yong; they were not to be numbred, most of them being yong and naked, and this he onely did, by casting up their Nests, which are always built in a great heap, of double bigness to the rest, most easily discerned, and then the old ones would come to look their yong, which he would snap up presently also: At another Season then *March*, which is their time of breeding, such success is not to be expected. In other times the best way is, if there be any Hedges near, to set the Gins or Traps there, for their ordinary roads are in such Hedges, and other places they cast up, are but of uncertain use; as when they intend forage for one time, though it may be that they minde the use of that passage no more at all. *Bellonius* advises to bury Moles in those places, whence you would drive the rest of that Vermine; and there may be somewhat in that remedy: For many living Bodies have a great dislike to, and antipathy against the putrified Bodies of their own kinde: Thus Worms, putrified at the Belly of a Childe outwardly, and the powder given inwardly, are

are esteemed as Medicines destructive to the Worm in the Belly, though the latter way is by some thought to breed more then it kills. Nay, in Vegetables 'tis agreed, That a yong Orchard will not thrive among the Roots of an old rotten Orchard; the reason whereof, some suppose to be the antipathy of the yong, against the old putrifying Roots; but of this effect, other reasons may be as probable.

There be some other remedies for the same annoyances, as, particularly, for the destruction of Fearn, the Author named gives this prescription: In the Spring, when the Fearn begins to grow a little above the Grass, while it is yong and tender, take a crooked Pole, or piece of Wood about six foot long coming in at one end like a Bow; or made like a blunt Sithe; with this strike off all the heads of the Fearn, as low as you can, even to the ground, if possible; do this the second or third time; and it proves generally a certain remedy. The reason, as I suppose, is the putrefaction of the Fearn, it being a very moist Muscilaginous Plant, by its own juice, and the moisture of the Earth, by which the very Roots themselves come to be corrupted, or else the deprivation of all the Buds that germinate from the Root, by cutting off the Sprouts so unseasonably.

For Ant-hills, to destroy the Insects, and take the hills down, this manner is prescribed; Divide the upper Turf into five or six parts, then take it down with a turfing Spade to the bottom of the Banck; the Turf being cut as thin as can be under the roots of the grass; then take out the Core of the Bank, that when the Turf is returned to its place, it may lie there lower somewhat than the surface of the Earth,

that

that the moisture, which will be a certain destruction of the Ants, may a little reside there: This must be done in *November, December,* or *January,* that the Roots of the Grass may the better take to the ground, before hot weather comes in the Spring.

Among Mr. *Speed's* notes, there are these Recipts, take red Herrings, and cutting them in pieces, burn the pieces on the Mole-hills; or you may put Garlick or Leeks in the Mouthes of their Hills, and the Moles will leave the ground. I have not tryed these ways, and therefore refer the Reader to his own tryal, belief, or doubt.

I had almost forgot to mention the change of Seed from grounds of a contrary nature, which by the experience of Husbandmen is found very advantagious, and is thought to prevent smootiness. 'Tis the custom in *Buckinghamshire,* for those of the Vale to buy their Seed from the Chiltern, on this account; and this experiment is found profitable in Wheat, Barley, Pease, and all Field Grains; and not so onely, but also in Garden Plants.

For the preserving early or late sowed Corn, or the same when it begins to corn in the Ear, from Crows, Rooks, or Jack-Daws, Mr. *Blith* has invented this Scare-Crow: You must, says he, kill a Crow or two, and take them into the Field where they haunt, and in the most obvious, plain, perspicuous places, make a great hole of two foot over, and about twenty Inches deep, on the highest ground in the Field, which hole must be stuck round about the edges with the longest Feathers; the bottom must be covered with the shortest, and some part of the Carkass; and that Turf or Earth that is digged out of the hole, being laid round upon a heap, you may stick round with

Feathers

Feathers also. One Crows Feathers will dress two or three holes, and about six or eight holes will serve for a Field of ten or twelve Acres. The Feathers will remain fresh a Moneth, unless store of Rain or Weather beat them much; and then (if needful) they must be renewed.

CHAP. VIII.

Of the Means of Improvement and best culture of such Plants or Flowers as are usually cultivated in Gardens or Orchards, and of the ways used for the removing of such annoyances as are commonly incident to them.

Num. 1. *Of the annoyances in general incident to Garden Plants.*

THe Politician speaks it to be a part of as great skill and prowess to defend a place already gotten, and to improve it to the benefit of the Prince and Inhabitants, as it was at the first to arrive at the Conquest; this is alike true in the Gardiners Province: It is no easie thing with him to raise a stock of choice Plants, by the several ways of propagation above mentioned, and as hard to preserve them, being propagated, from destruction by foreign and intestine violence. For either the sharpnesse of cold, the torridness of the Sun, Vermine, or other accident from without; or want of convenient and nourishable soyl of earth and water, and other Elements proportionable to the plant, will be such internal deficiencies, as to cause utter destruction: or the hastiness and premature, or on the contrary, the tardy and slow germination thereof will hinder its excellency; or weeds, or other vegetables, may grow up to its hinderance: and many other impediments there are, which with their several remedies, as they shall suggest themselves to my thoughts, I shal propose in the present Chapter, the last of this discourse.

N. 2. *Of defences for choice plants from cold.*

One great annoyance to all choice flowers and tender plants, arises from the violence of the Winter cold, the defence against which you shall have as far as I am able to give you, and can think of in the following directions.

Let those Bulbous Roots that are tender, such as the great double white Daffodill of *Constantinople*, and other fine Daffodills that come from hot Countries, the *Ornithagolum Arabicum*, purple Montain, Moly, &c. be planted in a large Tub or pot of earth and housed all the Winter, that so they may be defended from the frosts, or else, (which is the easier way) keep the Roots out of the ground every year from *September* after the leaves and stalkes are past untill *February*, in some dry but not hot or windy place

place; and then plant them in the ground under a South-wall, which are Mr. *Parkinsons* directions.

Alsoe the late Pine-aple Moly, the Civet Moly of *Mompelier*, the little hollow white *Asphodill*, which though its roots are not glandulous as to be capable of the last way, yet they are well preserved many yeares if by housing they shall be defended from the winter wett and cold.

Rose-bay Mirtles, the *Indian Gelsimines*, *Jucca Indica*, Orange trees, must be housed in the Winter, so likewise, the Cypresse, Bay, Piracantha, Mirtle, Pine-tree, Rose-bay with Spanish seed, or at the least must be cover'd with straw, or Ferne, or bean-hame, or such like thing layd upon crosse-sticks to bear it up from the plants till they are two or three yeares growth and fit to be removed to their places; *Arbutus*, or the Strawberry tree, Sea-Ragwort, the Pomegranate, and the Indian Figge require the same care.

Ferrarius commends a Garden house with Walls of thick mosse as good, and so without question it is, against the Winter cold and Summer heat.

Some defend their Mirtles, Pomegranates, and such other tender plants, either by houses made of straw like Bee-hives, or of boards (with inlets for the Sun by casements, or without them) Litter of Horse-stables being layd in very cold weather about the houses of defence.

It was a custome in *Italy*, to make such fences for Myrtles (especially when young) as appeares by *Virgills* Verse.

Dum teneras defendo a frigore Myrtos.

The Roots of the Marvaile of the World, Mr. *Park.* has preserved by art a Winter, two or three,

(so

(for they'l perish being let out in a garden, unlesse it be under a house side or such dry place) because many times the year not falling out kindely, the plants give no ripe seed, and so Gardiners would be to seek for seed to sow, and Roots to set, if this or the like art to keep them were not used: Tis thus, Within a while after the Frosts have taken the plants that the leaves wither and fall, dig up the Roots whole, and lay them in a dry place for three or foure dayes, that the superfluous moysture on the outside may be withered and dryed; which done, wrap them up severally in two or three browne papers, and lay them by in a box, chest, or tub, in some convenient place of the house all the winter time, where no wind or moist air may come unto them, and thus shall you have these Roots to spring afresh the next yeare, if you plant them in the beginning of *March*, as Mr. *P.* has by his own relation sufficiently tryed, but some have tryed to put them up in a barrell or firkin of sand and ashes, which also is good if the sand and ashes be throughly drye, but if it be any thing moist, or if they give again in the Winter, as it is usuall, they have found the moisture of the Roots, or of the sand, or both, to putrifie the Roots.

The same Author takes notice that tis one great hurt to Gilly-flowers in the Winter, and to all other herbs, to suffer the Snow to lye upon them any time after it is fallen; for it doth so chill them, that the Sun doth (though in Winter) scorch them up, shake therefore off your snow gently, not suffering it to lye on a day if you can; There is the like inconvenience from Frosts which corrupt the Roots, and cause them to rot and breake, for prevention,

take

take straw, or Litter of an horse stable, and lay some thereof about every Root of your Gilly-flowers, especially the best sorts, close unto them upon the ground, being carefull that none lye upon the green leaves, or as little as may be: Let it lie till *March* (with its winds) is past. The generall Remedy for these and all flowers, is to be covered with mats, which are removeable at pleasure. The choicest of all are put in pots and housed.

Num. 3. *Of shades requisite to sundry Plants, especially when young, for their defence from the Sun and Winde.*

All sorts of Carnations, Gilly-flowers, and Plants that are tender and yong, especially your *April* and *May* Seedlings, are to be preserved and defended from the violent heat of the Sun, and blasting Winds: I have seen whole Beds of divers sorts of young Seedlings, utterly burnt up at their first appearing, by the violence of two or three hot days. Nor do Seedlings onely require this, but all Plants that are not altogether wild, of how woody substance soever, that are newly growing, from cuttings, or parts without actual Roots.

Shades are commodious, if not absolutely necessary to many Plants, even when they are well rooted, as Bays, Lawrel, Savin, and most Wood-plants, a mixture of Shade and Sun to Straw-berries; so that the Lord *Bacon* wittily advises, to sprinkle a little Forrage-seed on the Strawberry-bed, for that the Straw-berries, under those Leaves, grow far more large then their fellows.

The best shades are made by thin well pruned Hedges,

Hedges drawn through the Garden or Nursery, or by Mats laid over them, and underpropt by a frame of light Poles: But all Seedlings, Flowers, or other Plants that are kept in Pots, are readily removed into convenient shade at pleasure.

Of watering.

Watering with water that has stood two or three days in the Sun, is absolutely necessary for all Stringy Roots that I know, at their first removals; and at any time, when any Trees or Plants are weak, by reason of Drought: All manner of Layers must be specially regarded for matter of watering; and those Plants which are to be propagated by the circumposition of a Basket of Mould, (to make Dwarf Plants, as they call them) are specially to be watered in dry times: All maner of Gourds, Melons, Cucumbers, even in ordinary weather, require this help, although already firmly rooted.

But there is this difference in Plants, Those that require an hungry ground, shall well be content with thin water Sun'd: But Kitchin ground is best improved by fat water, wherein Ordure has been washed.

And some caution is to be had, that by too much water you do not chill or over-glut the ground, often and little is the best use, and in the Spring and Autumn when Frosts are feared, 'tis better watering in the Morning then at Night; in Summer, the Night I esteem the better Season.

There is a pretty way of watering choice Plants, by wetting a streiner, and so letting one end of it hang over a Vessel of water, which will draw up the

moisture

moisture from the Bason, and let it gently fall down the streiner to the Root of the Plant.

N. 4. *Examples of the best Culture of Hops, and ways of ordering them after they are first set, taken out of Mr. Blith.*

When, says he, your Hops are grown two foot high, binde up with a Rush or Grass, your springs to the Poles, as doth not of it self, winding them as oft about the Poles as you can, and winde them according to the course of the Sun, but not when the dew is upon them; your Rushes lying in the Sun will toughen, says he, but surely better in the shade.

And now you must begin to make your Hills, and for that purpose get a strong Hoe, of a good broad bit, and cut or hoe up all the Grass in the borders between your Hills, and therewith make your Hills, with a little of your Mould with them, but not with strong Weeds; and the more your Hills are raised, the better, the larger, and stronger grows your Foot, and bigger will be your fruit; and from this time you must be painful in your Garden, and be ever and anon, till the time of gathering, in raising your Hills, and clearing your Ground from Weeds.

In the first year suppress not one Cyon, but suffer them all to climb up the Poles, for should you bury the Springs of any of your Roots, it would die, so that the more Poles are required to nourish the Spring. But after the first year, you must not suffer above two or three stalks to grow up to one Pole, but pull down and bury all the rest. Yet you may let them grow four or five foot long, and then choose out the best for use. As soon as your Pole is set, you may
make

make a circle how broad your Hill shall be, and then hollow it, that it may receive the moisture, and not long after, proceed to the building of your Hills.

And where you began, or where your Hops are highest, there begin again, and pare again, and lay them to your Hops, but lay the out circle highest to receive moisture; be alway paring up, and laying it to the heap, and that with some Mould, until the heap comes to be near a yard high, but the first year make it not too high; and as you pass through your Garden, have a forked Wand in your hand to help the Hops that hang not right.

Now these Hills must the next year be pulled down, and dressed again every year. Some, when their Hop binde is eleven or twelve foot high, break off the tops, which is better then they that have their Poles so long as the Hop runs: But if that your Hop, by the midst of *July,* attain not to the top of your Pole, then break off the top of the same Hop, for the rest of the time will nourish the branches, which otherwise will loose all, it being no advantage in running up, to the stok or increase of the Hop.

In *April*, help every Hill with a handful or two of good Earth, when the Hop is wound about the Pole; but in *March* you will finde, unless it hath been tilled, all Weeds; but if you have pull'd down your Hills, and laid your ground, as it were, level, it will serve to maintain your Hills for ever; but if you have not pulled down your Hills, you should, with your Hoe, as it were, undermine them round, till you come near the Principal Root, and take the upper or yonger Roots in your hand, and discerning where the new Roots grow out of the old sets, of which be careful, but spare not the other; but in the first year, uncover no more but the tops of the old sets, but cut no Roots before the end of *March*, or beginning of *April*. The first year of dressing, you must cut off all such as grew the year before, within one inch of the same: and every year after, cut them as close to the old Roots; those that grow downward, are not to be cut, they be those that grow outward, which will incumber your Garden, the difference between old and new easily appears; you will finde your old sets not increased in length, but a little in bigness, and in few years, all your sets will be grown into one; and by the colour also, the main Root being red, the other white; but if this be not early done, then they will not be perceived: And if your sets be small, and placed in good ground, the Hill well maintained, the new Roots will be greater then the old; if they grow to wilde Hops, the stalks will wax red, pluck them up and plant new in their places.

N.4. Mr. Parkinsons *way of ordering the seedlings of Tulips grown.*

After the Tulip seed is sowne, the first yeares springing bringeth leaves little bigger then the ordinary grasse leaves; The second yeare bigger, and so by degrees, every year bigger then other. The leaves of the præcoces, while they be young, may be discerned from the Media's, by this note which I have observed, The leaves of them do stand above ground, shewing the small foot-stalkes whereby every leafe doth stand; but the leaves of the Media's or Serotines do never wholy appear out of the ground, but the

the lower part which is broad abideth under the upper face of the Earth.

Those Tulips now growing to be three yeares old (yet some at the second yeare, if the ground and aire be correspondent) are to be taken up out of the ground (wherein you shall find they have run deep) and be new planted after they have been a little dry'd and cleansed either in the same or another ground, again placing them reasonable neere one to another, according to their greatnesse, which being planted and covered over with earth again, of about an inch or two thicknesse, may be left untaken up again two yeares longer, if you will, or else removed every yeare after, as you please, and thus by transplanting them in their due season (which is still at the end of *July*, or at the beginning of *August*, or thereabouts) you shall according to the seed and soyle, have some come to bearing in the first year after their flowering. some have had them in the fourth: (but that hath been but few and none of the best, or in a rich ground) some in the sixth and seventh, and some peradventure not untill the eighth or tenth yeare. But remember that as the roots grow greater that in the planting you give them the more roome to be distant one from another, or else the one will hinder (if not rot) the other.

The seed of the Precoces do not thrive and come forward so fast as the Media's or Serotines, nor do give any off-sets in their running down, as the Media's do, which usually leave a small Root at the head of the other that is run down every yeare; and besides are more tender and require more care and attendance then Media's, and therefore they are the more respected,

This

This is a generall Rule in all Tulips, that all the while they beare bud or leafe, they will not beare flower, whether they be seedlings, or the off-sets of elder Roots, or the Roots themselves, that have heretofore borne flowers; but when they beare a second leafe breaking out of the first, it is a certain signe that it will then bear a flower, unlesse some casualty hinder it; as Frost or Raine, to spoile or nip the bud, or other untimely accident befall it.

To set or plant the best and bearing Tulips some what deeper then other Roots, I hold it the best way. For if the ground be either cold or lye too openly in the cold Northern aire, they will be the better defended therein, and not suffer the frost or cold to peirce them so soon, for the deep frosts and snowes do pinch the Precoces cheifly, if they be too neer the uppermost crust of the earth, and therefore many with good successe cover over their ground before winter with either fresh or old rotten dung, and that will marvellously preserve them. The like course you may hold with seedlings, to cause them to come on the forwarder, so that it be after the first yeares sowing and not till then.

To remove Tulips after they have shot forth their Fibres or small springs which grow under the greater round Roots (that is from *September* untill they be in flower) is very dangerous, for by removing them when they have taken fast hold in the ground, you do hinder them in the bearing out their flower, and besides put them in hazard to perish, at least to be put back from bearing a while after, as often I have proved by experience, but when they are now risen to flower, and so for any time after, you may safely take them up if you will, and remove them

without

without danger, if you have any good regard to them unlesse it be a young bearing Root, which you shall in so doing much hinder, because it is yet tender by reason it beareth now the first flower, but all Tulip Roots when their stalks and leaves are dry, may most safely then be taken out of the ground, and be so kept (so that they lye in a drye, and not in a moist place) for six moneths without any great harme, yea I have known them that have had them nine moneths out of the ground, and have done reasonable well, but this you must understand withall, that they have not beene young but elder Roots, and have beene orderly taken up and preserved; the dryer you keep a Tulip Root the better, so as you let it not lye in the Sun or the Wind, which will pierce and spoile it.

Num. 5. *Of annoyance by Plants growing too thick and neer together, and of the remedy thereof, and improvement by pruning Trees, and setting them at great distances; plucking off the yong Germens of Garden-flowers, to make the rest more fair; of the sizing of Turneps, Carrots, Parsneps; of Weeding.*

There is no greater hindrance to the growth and thriving of all Vegetables, than to be so crowded together, that their Roots, Branches and Leaves, interfere one with another; and therefore in all Orchard and Garden-plants, whose Fruit and Flowers you require fair, and whose growth you would have considerable, provide that they keep their distances: Apple-Trees, Pear-Trees, Plum-Trees, Cherries, and other Plants, are of diverse statures, both in regard

gard of one another, and of their own kinde: Some Apple-Trees grow to much greater growth than some other, Pears to a greater growth then Apples, so that it is hard to appoint a certain distance for Trees in an Orchard, twenty Foot is space little enough for Standards of common Apples or Pears; but a certain rule is, to provide that one Tree shade not another, and therefore let the lowest Trees, if you intend to make the most of your ground, be set South, and the highest Pear-trees stand to the North; for should the higher Trees stand South, they would cast their shade over the rest of the Orchard.

This Doctrine of setting Trees at such distances, the Husbandman hates, for two reasons; one is, Because it takes too much of his pasture from his Cattle; and the other is, That by this means he can have but little Fruit in his Orchard for many years: Therefore to gratifie his covetousness, I shall propose him this practicable way of following and prosecuting my intention to the utmost profit, without putting him to the mentioned grievances. For first, I shall order that he plant his Orchard full of Trees, within three yards distance one of another, or somewhat nearer, if he please; these shall bear him after a year or two, as many apples as a well-grown Orchard usually carries: then let him set this ground to a gardiner, that it may be digged and dunged seasonably, to bring Kitchin Plants; for from this Culture the Trees will receive great advantage. When the Trees are big enough, with the defence of a strong stake, and some Bushes, to be secured from Cattle, let him transplant them into Pastures of the best Soyl, where they may stand at great distances to be shelter to Cattle, and no prejudice to the Grass: One Tree at such distance,

shall

shall bear as much as ten in some Orchards, and thus continue removing, as your Trees grow big enough. I count five or six inches about to be a good Size, the bigger they are, the more care must be taken in their removal, that the Root be transplanted entire as may be, without much dis-branching it, or cutting away the spurs. And it is convenient, that in the heat of the first Summer, wet Straw be laid upon the ground about the Root.

If you have no pasture to transplant into, sell your Trees to those that have, or set your Standards of strong Trees at twenty foot distance, and fill up the rest of the ground with Kentish Codlings; Nurse Gardens, Burts, which are cheap Plants, being propagated by Suckers, or with dwarf Trees, made by Circumposition, which may be cut down when the other Orchard thickens too much, and in the mean time are very plentiful bearers.

Pruning Trees is used likewise chiefly to this intent, that the Rays of the Sun may have passage to all parts of the Tree, so that 'tis a good way for the Pruner to look upward from the North side of the Tree, upon the South and East, and to cut off, or rather make thin, such boughs which he findes so thick as to obstruct the Sun: All Boughs likewise that gall others, and that are actually dead; providing always, that the Boughs taken off be as little as may be, though the more in number, that so the sap may make up the Bark, and the Tree be not decayed by lopping of the greater stems: Which is very perversly done by most Gardiners, who think that to Prune a Tree, is to cut off the lower Boughs bigger or less, because they see small watery Fruit grow on them; whereas if the Sun was let in upon them,

them, their Fruit would be rather more, than lesse forward, than that which grows in the middle of the Tree: I count it general, that the under-Boughs ought never to be cut off, but when you have respect to grass Roots, or other Garden-stuff, which grows under the Trees, or for the security of the Trees from the browsing of Cattle, so that to bare the Trunck of the Tree, for four, five, or six yards, as some doe, and nourish it to no profit, but to bear and carry up the head to another Region, that Rooks may the better build therein, is a common folly, and ridiculous, if well considered.

And for lopping off great Boughs, I may here adde an observation touching Elms, which is, That if the top of an Elm of any bignefs be cut off, the rot will immediately begin there, and by wet, and other accidents, run downward, and cause that hallowness which is ordinarily seen in Trees of this kinde.

Another Rule of pruning, is, That the Gardiner never cut off those Boughs which are set and adopted for bearing, which is easily known for Roses particularly: Rasps and Vines always bear upon a fresh sprout, shot forth the same Spring, so that the more you prune a Rose, Rasp, or Vine, the more fresh sprouts of that Springs growth are emitted, and the more such sprouts, the greater number of Roses, Rasps and Grapes succeed, unlefs some particular accident destroys them. Many Fruits bear from the shoots of the antecedent Spring, as the generallity of Apples, Pears, Peaches, Nectarins, Aprecots: Many seem to grow from Wood of longer growth, but in that a man may be easily mistaken, because a very little, and a Spring of scarce discernable growth, may be enough to serve as a foundation to the pedal of the

Blossom or Fruit, which standing on the old Wood, it may be thought that the pedal or stalk of the Fruit, stands immediately on the Wood, and that there was no Spring interceding. Sometimes the Blossoms of the same Tree, stand both on the Wood of the present and antecedent Spring, as it is frequently seen in Kentish Codlings, Nurse Gardens, great bearing Cherries. But where ever the Blossoms are, and there are many Buds fitted and prepared for bearing, they are discerned by the skilful Gardiner, and may be seen by any person, for those are more full in their shutting up than other Buds are, and stand not so close made to the stem of the Branch whereon they grow, and contain more small leaves in their Body then other Buds, being, as I apprehend, the actual rudiment of the ensuing Blossom: Such Boughs therefore, whereon plenty of these full made Buds, or inchoate Blossoms are seen, the Gardiner spares, if he is wise, for the present year, and (where he may) prunes off such whereon he sees no such propension to fruitfulness.

The fairness and largeness of Flowers and Fruits are very much augmented, by preventing the running up a multitude of Stalks from the same Root: The Gardiner observes this precisely in his Carnations and Gilly-flowers, not suffering above one, two, or three Spindles upon such Roots or Stools where he intends a greater fulness and largeness in the Flowers; and in Anemones the observation is, That if any of the Latifolia's bring a single Flower, on the same Root with the double, then the cause usually is, the standing of too many Eyes or Germens, and their depending from the same Root; and the remedy in like manner, nothing else but the taking off those Off-sets or Suckers,
and

and parting them from the principal Root, which otherwise is robbed of that matter which might raise in each Flower, both fairness and multiplicity of leaves.

Shrubs likewise that bear either Fruits or Flowers, are to be governed in like manner; Goos-berries and Currans degenerate to smalness, or bear not at all, without this care and provision, that the Suckers be taken away: This observance is likewise absolutely necessary to Damask Roses, for when they grow up to thick Bushes they scarce bear, whereas being kept to grow in one single great stem, being orderly cut, and not growing in the shade, they bear exceedingly.

For Vines, it is a Proverb, make your Vine poor, and it will make you rich: The fewer principal Stems are left, the more it bears, and the reason is, because the Grapes are borne upon shoots of the same Spring; and those shoots then most plentifully arise, when the head of the Vine, in proportion to the Roots, is least, as 'tis seen in all Trees, which shoot out more immediately after their heads are lopt, than any other year. Pompions follow the nature of Vines, and as two or three stems is enough for the Vine, so two or three runners, and no more, ought to be permitted by him that intends the greatest fairness of this fruit.

It may be proper enough here to speak of Weeding and Sising: The latter operation is, the plucking up Roots or Plants that are of use in themselves, but offensive to others in the same Beds, by reason of their nearness: Thus Turneps are howed up when they stand within a foot distance each of other; for it is best, when at their full growth their leaves touch not one

one another: Carrots are plucked up, when they are an inch Diameter at the head, for then they are of use, or sooner, if the thickness of their standing require it; and this is general for all Roots, Parsneps, Radish, Skirrets, that grow by Seed: Some sow (as I mentioned above) Parsneps, Carrots, Radish, and Sallad Herbs in the same Ped, first Sising out the Sallad Herbs and Radish, then the Carrots as they grow, leaving the Parsneps till Winter, by which means their ground is always full, yet by reason of the Sising in due times, never over-burthened.

The culture of Straw-berries requires somewhat like sizing, (*viz.*) The cutting off immediately after bearing the spires and strings, which would multiply unto too many Roots and Branches, to have plenty of fair Straw-berries: Nor is this once onely to be done, but as often as they spring anew, so often are they to be taken off, until the time of the Blossoms draws on; I have seen some that were not over curious to tear off the strings by harrowing up and down their Beds of Straw-berries with an Iron Rake.

Some make a question, Whether Plants of the same kinde, by reason of a supposal that they require the same parts for nourishment; or Weeds and Grasses, by their too great vicinity, may create more annoyance to their Neighbors, I decide not the question, nor can reconcile the Gardiner to Weeds, whilest he findes his strongest Plants destroyed by them: I have seen many Trees in a well grown Nursary, spoiled by the Grass that grew amidst them; and as I remember, the very Bark of the Trees themselves was rotted, by a dew cast upon them from the Grass: I have likewise observed, a strongly grown Quickset of White Thorn, to have been destroyed by

Alexan-

Alexanders, which it is at the Readers choice to account as a Weed or cultivated Plant.

The time of pruning generally is the dead of Winter, for such Plants as consist of a Woody substance: Pompions are deprived of their superfluous creepers, and other Gourds likewise, at their first time of springing and divarication of their Branches. The season of pruning for acceleration of ripeness, is when the fruit is made, and begins to grow to some bigness, as generally they are, about Mid-Summer: Some have a third time of pruning Wall-Fruit, *viz.* at the time when the Fruit is taken off, as they do Roses likewise, when the Flowers are newly gone.

To cut the Branches or Sprigs of a Flower or Tree quite off, cannot properly be called pruning, yet sometimes it proves an useful operation for such Plants as are stunted, as they call it, in their growth; Trees that are crooked, or have been bitten with Cattle, or are grown old : Thus Wood-men count it best to cut those Stools of under-Wood down to the Root, that it may begin to shoot afresh, that have been much browsed by Cattle; and cut down their hedges to the Roots when they grow old and Mossy.

Gardiners likewise, if by reason of a sharp Winter their Anemone's are pinched with cold, and starved, let them not immediately run to flower, but cut off the first Springs to the ground, that in a better Season they may lay a stronger Foundation for the bearing of fuller and fairer Flowers.

Num.

N. 6. *Of Pismires, Earwigs, Canker and rottenness in choice Plants, Catterpillars, Mossiness, Bark-binding, Bursting of Gilly-flowers.*

There are many other annoyances to Vegetables, and generally sooner reckoned than remedyed, a word or two I shall speak of as many of them as come into my minde: Pismires, especially those of the black kinde, are exceeding troublesome in some Gardens, for they climb the highest Trees, and spoil the Fruit, are commonly esteemed remediless. *Bellonius*, who took exceeding pains for improvement by Vegetables, commends the decoction of Broath made of any sort of Spurge, as very efficacious for this purpose: Some draw them to one place, by burying Carryon where they most resort, and then scalld them with seething liquor.

To divers choice Flowers, but Carnations and Gilly-flowers especially, Ear-wigs are a great annoyance: Mr. *P*'s way of setting Beasts Hoofs among the Flowers, upon sticks, to take them, is used of every Body here, and generally lik'd: Some that set their Flowers in Pots, set the Pots in Earthen Plates, with double Verges, containing water, or water mingled with soot in the outward verge, to drown the Vermine that shall attempt the pots, and rain water in the second, which may pass through the holes of the pots to water the earth therein contained.

The rottenness and hollowness, that through age & too much moisture bulbous and tuberous roots, and the best Anemones especially, are subject too, is thus provided for; the disease must be laid open, and the rottenness cut out so, that in the root there be no capacity

ty left to hold water, which I have often mentioned to be a great Enemy both to them and Tulips. *Ferrarius*, and some others, prescribe Plaisters of Rosin, Turpentine, and Wax, to apply to the Cicatrices of the wounded Root, which notwithstanding, I have no great regard for. The same Author says, that in moist Winters Anemones do best in pots, in dry, better in beds: With us they are seldom potted, but the borders for these Plants are usually laid on pretty high ridges, as Husbandmen lay their Corn Land in deep and moist ground, to prevent the mischiefs that usually happens by too much wet.

Mr. *Parkinson* says, That if you perceive that your Gilly-flower leaves change any of their Natural fresh colour, and turn yellowish, or begin to wither in any part, it is a sure sign that the Root is infected with some canker or rottenness, which will soon shew it self in all the rest of its branches, and therefore betime, (else 'tis in vain) advises that you cover all, or most of the Branches, with fresh Earth, or else take the fairest slips from it, or according to Art lay it: This way of Mr. *P.* may be applyed unto other Vegetables.

I know no better way to destroy Catterpillars, Palmer worms, and other Vermine of that kinde, then by crushing their Eggs; as soon as they are laid upon the leaf by the Fly, some brush them off with wet cloathes: 'Tis observed, that the little Fly that usually blows upon the Cabbage, chooses such Plants as are yongest, and especially those that were raised in hot beds, or endured least of cold in the Winter preceding.

Mossiness of Trees, comes generally either from the barrenness or coldness of the ground, and therefore

fore I count it vain to attempt the removal of it, without taking away the cause, and making the ground better; which being done, it will be proper enough to rub down the Trees in a wet day with an hair cloath.

Trees likewise are sometimes Bark-bound, especially such, the grain of whose Bark runs round the body of the Tree, as in Cherry-trees, and not straight upward, according to the grain of the Tree, as in Apples, Pears, &c.

For the Bark is not generally, as I suppose, nourished by apposition of a new rinde to it, as the substance of the Tree is, but by interposition of particles, amidst the particles of the rinde already made, which if it be so hard as not to admit other Particles for its enlargement, there can be no new addition of a new coat of wood, which ought to accrue every year to the Tree, for there will be no space wherein the sap may ascend, which is to be hardned into such new wood, unless by renting the whole coat of Bark, which sometimes happens.

The remedy for this disease, both in Cherry-trees, and other Trees, those chiefly whose Barks are hardned and grown crusty by long standing in shadowy places or barren ground, is, that the year after their removal, or upon addition of better soil in streight grained Barks; and without either removal or addition of soil in Cherry-trees, and other cross grained Barks, or in any Trees whose Barks rend of their own accord, the Barks be slit from the top of the Tree to the bottom of the stock, and that according to the bigness of the Tree, in one, two, or three places: This is a Chyrurgical remedy that never fails, and is easily performed.

Carna-

Carnations and Gilly-flowers, happen to be often deformed, especially those which are of the largest sorts, by bursting the Calyx, Cellar, or Case wherein they are set, and the usual remedy is, to inlarge the five incisions proportionably, by cutting them deeper with a Knife; or to steep ordinary Beans in Water, and then slipping off the outward coat of the Bean, to put it (the end being taken off) upon the head of the Carnation, which will keep the five lips together, and preserve the Flowers from breaking; nor will these Hoops, made of the coats of Beans, shrink with the heat of the Sun, as those made of the rind of Willow, slipped off for the same purpose, usually doe: One Bean is long enough to make two hoops, for they need not be above a quarter of an inch in breadth.

Num. 7. *Of improvement and melioration of divers Sallad Herbs, by blanching or whiting, from the French Gardiner, and Mr. P's Observations.*

The Lettuce-Royal, being, upon removal, set at a foot or more distance, when you perceive that the Plants have covered all the ground, then in some fair day, and when the morning due is vanisht, you shall tie them in two or three several places one above another, which you may do with any long straw, or raw hemp, and this at several times, (*viz.*) Not promiscuously, as they stand, but choosing the fairest Plants first, to give room and air to the more feeble, and by this means they will last you the longer: The first being blanched, and ready before the other are fit to binde.

If you would blanch them with more expedition, you may cover every Plant with a small earthen pot, fashion'd like a Goldsmiths Crucible, and then lay some hot soyl upon them, and they will quickly become white.

Concerning Succories, Thus,

There are several kindes of Garden Succories, different in leaf and bigness, but resembling in taste, and which are to be ordered alike.

Sow it in the Spring upon the Borders, and when it has six leaves, replant it in rich ground, about eighteen inches distance, paring them at the tops: when they are grown so large, as to cover the ground, tye them up, as I instructed you before, where I treated of the Roman Lettuce; not to binde them up by handfuls, as they grow promiscuously, but the strongest and forwardest first, letting the other fortifie.

There is yet another fashion of blanching it: In the great heats, when instead of heading, you perceive it would run to Seed, hollow the Earth at the one end of the Plant, and couch it down without violating any of the leaves, and so cover it, leaving out onely the tops and extremity of the leaves, and thus it will become white in a little time, and be hindred from running to Seed.

Those who are very curious, binde the leaves gently, before they interre them, to keep out the Grit from entring between them, which is very troublesome to wash out, when you would dress it.

Remember to couch them all at one side, one upon another, as they grew being planted, beginning with that which is nearest the end of the Bed, and continuing to lay them, the second upon the first, and the third

third upon the second, till you have finished all the Ranges.

I finde likewise two other manners of blanching them for the Winter; the first is, at the first Frosts, that you tie them after the ordinary way, and then at the end of eight or ten days, plucking them up, couch them in the Bed where you rais'd them from Seeds, making a small Trench cross the Bed the height of your Plant, which will be about eight Inches, beginning at one end. In this you shall range your Plants side by side, so as they may gently touch, and a little shelving; this done, cover them with small rotten dung of the same bed: Then make another furrow for a second range, in which order, lay your plants as before, continuing this order till you have finish'd: And last of all, cover the whole Bed four Fingers thick, with hot soyl fresh drawn out of the Stable, and in a short time they will be blanched. If you will afterwards cover the bed with some Mats placed aslant, like the ridge of a House, to preserve them from the Rain, they will last a very long time without rotting: When you would have any of them for use, begin at the last which you buried, and taking them as they come, draw them out of the range, and break off what you shall finde rotten upon the place, or that which has contracted any blackness from the dung, before you put it into your Basket for the Kitchen.

A second manner of preserving it, is, to interre it, as before, in furrows of Sand in the Cellar, placing the Root upmost, least the Sand run in between the leaves, and you finde it in the dish when they serve it. You need not here bestow any dung upon them, it is sufficient that the Sand cover the Plant four fingers

gers high; and when you take it out for use, before you dress it, shake it well, the Root upmost, that all the Sand may fall out from the Leaves. Take them likewise as they happen to lie in the Ranges.

His directions for blanching Endive, are, that you cover it onely with reasonable warm dung, and drawing it out at the first appearance of Frost, that you keep it under Sand in your Cellar, as you do other Roots, but first it must be almost white of it self.

The whiting of Endive, Mr. *Parkinson* commends, when done in another manner: After, says he, that they are grown to some reasonable greatness, but in any case before they shoot out a stalk in the midst for Seed, take them up, and the Roots being cut away, lay them to wither for three or four hours, and then bury them in the Sand, so as none of them may lie one upon another, or if you can, touch one another, which by this means will change whitish, and thereby become very tender, and is a Sallot for Autumn and Winter. Fennel is whited by some in the same manner, for the same use.

To procure the Chard of the Artichocks (which is that which groweth from the Roots of old Plants) you shall make use of the old Stems which you do not account of. For it will be fit to renew your whole Plantation of the Artichocks every five years, because the Plant impoverishes the Earth, and produces but small fruit.

The first Fruits gathered, you shall pare the Plant within half a foot of the ground, and cut off the stem as low as you can possible; and thus you will have lusty slips, which grown about a yard high, you shall binde up with a wreath of long Straw, but not too close,

close, and then environ them with dung to blanch them.

Thus you may leave them till the great Frosts, before you gather them, and then reserve them for your use in some Cellar, or other place less cold.

N. *Of Acceleration and Retardation of Plants, in respect to their Germination and maturity.*

Acceleration of Plants in their Germination and Maturity, is ranked, by the Lord *Verulam*, among the *Magnalia Naturæ*, and is an operation that all Artists can do something in: though I know not any that arrive to the performance of those grand proposals of some Writers, as that of raising Sallads within an hour or two, whiles a Joynt of Mutton is rosting: The late King of *France*, has been reported to have known a secret process that would produce this effect, and to have esteemed it at a high rate: Cichory was the Seed, as I was informed by Monsieur *Gissonius*, which he was wont to raise so soon into his most fam'd Sallad.

I have tryed divers of the Experiments proposed for procuring those wonderful speedy Germinations, and that by long infusions in Milk, strong Muck-water, and sometimes have added unquenched Lime unto the infusions, according to the Experiments set down by a late Writer, who asserts, that by these usuages, Beans, Pease and Parsly Seed would grow up in few hours, and can onely give the Reader this fruit of my pains, that without any further tryal, he may from my experience be ascertained, that the advantage in acceleration is exceeding inconsiderable by any of these means. It was, by my tryal, found much less

than

than I imagined could have been by any infusion, for none of the Seeds (of which I tryed many sorts) came up the first three or four days; and except Radish, none came up in a fortnights time, though they were sown in *August*, and watered.

I have likewise tryed the Experiment of Ashes of Moss: First, burning a great quantity of Moss to ashes, and then taking some of the richest Garden mould I could procure from a rotten hot Bed, and mixing it with the ashes, I moistned it with exceeding good Muck-water several times, and let it as often dry in the Sun; this I did in glazed pans, that the Salt might not be washed from the Earth; then I sowed Seeds, some unsteeped, some steeped, and in the beginning of *September* set the Pans upon the Leads of an House: But in effect, the Sallad sprang not up that day, nor many days after.

The next day I set into some of the same kinde of Soyl, made up of Moss-ashes and dung, watered as above, divers Seeds steeped in Spirit of Urine alone, Spirit of Urine with water mixed, Spirit of Urine mixt with phlegm of Elder-berries, all without success, though I set them in a Pan of Earth over a gentle fire, to speed the Germination: Formerly I have seen Spirit of Nitre tryed, but to no purpose; some speak of working these suddain Germinations by somewhat made of Salt, Spirit, and Oyl, chymically united into one Body, which when they shall discover unto us, or otherwise make us possessors of, we shall have a better opinion of the related experiment.

As to ordinary Acceleration, hot Beds are the most general and catholick help, and certainly forward Germination much: For Cabbage-seed sown in the Spring on a hot Bed, I have seen, to bring Plants that have

have in their growth and bigness overtaken such that were re-planted before the antecedent Winter, and so were in the ground, at the least, half a year before them; and that in the same sort of Soyl. It is certainly true, that the Germination will be the more quick, the hotter the weather is; and the larger the bed of Dung is made, and the more it is helped by the reflection of Brick Walls, or other like advantages: The manner to make these hot Beds, is mentioned in the first Chapter, and their use there described.

Mr. *Speed*, *Cap*. 14. Of *Musk-melons*, Gives us from the testimony of two Noble Men, this advertisement: The way, says he, to have as good Muskmelons as any are in *Italy*, without the unwholesom use of the Muck-Beds here in *London*, is confirmed by the Earl of *Dorset*. Plant them under a Wall, Pale, or Hedge, on the Sunny side, with very good Mould purposely prepared, and underneath the mould lay a quantity of fresh Early-straw, and by this easie means, using the seasonable covertures and necessary furtherance, you may attain to your uttermost desire, without any further trouble. But if you do discern the Straw to make the Earth too hot, thrust in a Stake through the mould to the straw, that the vapor and heat may evaporate and pass forth.

For Acceleration of maturity in all Wall-fruits, the practice of Midsummer pruning is every where almost observed, which is, the cutting off all parts of the shoots that are grown out far beyond the Fruit, and do otherwise take away both the sap that might advantage the Fruit, and the benefit of the Sun likewise: This operation in Vines is called gelding, and is usually transferred to Pompions, Musk-melons, and

Cucum-

Cucumbers, and like Fruits, to accelerate their ripeness: The Joynt beyond the last Cluster or Gourd, is the place where the Creepers or Shoots are to be nipt off in Vines or Gourds: In other Wall-Fruit the Gardiner clips them at a convenient distance from the Wall, so as not to take away all the shade from the Fruit, which in some proportion is necessary that the Fruit be not dryed up, and burnt upon the Tree by the Torrid heat of the Mid-summer Sun, in such places where his rays are reflected from a Wall or Floor, or both.

'Tis also observed that in Wall-Fruit, or any other that requires a reflected heat, in order to the ripening of the Fruit; the lower the Boughs are spread, the sooner the Fruit ripens on a Wall: And in Standards, the lower and nearer the Earth any Plant is kept, the better shall it ripen, by reason of the reflection made from the surface of the Earth; which if be bare from Weeds, is equal to the reflection from some Walls. In *France*, Vines have no other reflection but this, being tyed to stakes, and not suffered to grow above a yard high; and in many places of *England* this onely advantage, without Walls, brings Grapes to that maturity which is ordinary in our Island.

The twisting of the stalks, whereby the Bunches of Grapes are joyned to the body of the Vine, done at such time when the Grape is come to its full bigness, is practiced by some for the accelerating maturity; and it may be, that by this twisting, the Channels, that might otherwise carry more crude Sap into the Grape, being broken, the heat of the Sun may more speedily reduce that which is already possessed by the Grape into sweetness, then if sowre and undigested Juice were still supplyed from the Vine.

Retardation, or hindring Plants from running to Seed, is likewise of use for the preservation of the Root and Leaf; for there are many Plants, whose last endeavor being to bear Seed, presently die in all parts of them assoon as the Seed is perfected.

Of this kinde are your best Carnations and Gilly-flowers, the hope of whose continuation is onely by those Slips that are not like to bring Seed the present year; to this kinde also belong divers Herbs, such as are Parsely, Scurvy-grass, &c. The Spindles therefore of all such are timely to be cut off, the younger the better, in choice Plants, for fear of killing the Root; and hereby plenty of Branches and Off-sets, or side-Plants, will arise from the old Stem, Stool or Root. Nay, 'tis observed by our Gardeners, as likewise by *Ferrarius*, in his Chapter of the culture of Tulips, That if those Flowers are suffered to grow to Seed, the Bulb thereby is certainly much emaciated, and sometimes utterly perisheth; and therefore, on all hands it is counted good to gather Tulips as soon as may be.

Some of the ways of Retardation are generally known, as particularly the experiment of plucking off Rose-Buds as often as they spring, until the time you intend they shall proceed to flower; or the making the Branches of the Rose Tree bare of Shoots once or twice in the Spring, for this purpose, are not unfrequently practiced. And I have been informed by a Person of Credit, that at *Bristol* he saw Raspes sold for four pence the quart at *Michaelmas*, which were thus retarded, by setting the Plants late in moist ground the same year: All which ways, I suppose, may well be transferred to other Plants of like nature, and this last way is not so common. I have before mentioned its use for the retardation of the Flowers of Anemonies.

There is some use of retardation to all such Plants which so prematurely blossom, that they be subject to blasting by Spring-Frosts; I know nothing used to prevent this annoyance, but the opening of the Root, and suffering the Snow, and Snow-water, to lie thereon and chil the ground; but of the benefit or danger of this remedy, I have no experience.

Num. 8. *Of melioration by Richness, or other convenient Minera in the Soyl, for the feeding and better nourishment of several Plants: Of artificial Begs, and the change of Seed, as a means to bring fair Flowers: Of Exossation of Fruit, or making it grow without Stones.*

The Lord *Verulam* reckons up the making of rich composts for the Earth, among the *Magnalia Naturæ*, and most advantagious projects for the use of Man; which richness, if the modern Hypothesis of Chymists be right, consists in good proportions of salt Spirit, and Oyl; which are principles generally deficient in barren places: Dry Earth, and cold crude water, or these two mixt together, every where abounding: I say, good proportions, because it is most certain, that no Vegetable will grow in too great abundance of Salt or Spirit, or other violently hot and corrosive matter: Sut and Pidgeons-dung abound much with volatile Salt; and I have this year, upon a cold moist Clay, seen excellent advantage to the Grass thereby, it being onely strewed thin on the Grass before the Spring, but of the two, the Sut was best: upon a dry Sand I should not have expected the
like

like improvement by its mixture, and in these composts themselves by reason of abundance of salt, without due proportions of other principles mixt, nothing will grow, for there is no fermentation without mixture of contrary parts or Elements; and all dunging is in order to fermentation: Hence *Columella* commends Pidgeon-dung, because, says he, *Præ cæteris terram facit fermentare,* the earth generally abounding in its own nature, with coldness & moisture, so that the richness in Salt or Spirit, tempers a Soyl well, which is deficient in these principles, for those Vegetables that require in the ground so sprightful a Fermentation. For divers states of ground, and various Fermentations are required to different Plants, nor can any one Soyl indifferently and equally agree with them all according to that of *Virgil.*

> *Nec vero terræ ferre omnes omnia possunt,*
> *Fluminibus salices, crassisq; paludibus alni*
> *Nascuntur; steriles saxosis montibus orni,*
> *Littora myrtetis latissima: denique apertos*
> *Bacchus amat colles, Aquilonem & frigora taxi.*
> *Aspice & extremis domitum, cultoribus orbem*
> *Eoasq; domos Arabum, pictosq; Gelonos*
> *Divisæ arboribus patriæ: sola India nigrum*
> *Fert ebenum, solis est thurea virga sabæis, &c.*

All Grounds can't all things bear: The Alder Tree
Grows in thick Fens; with Sallows Brooks agree.
Ash craggy Mountains: Shores sweet Myrtle fills,
And lastly, *Bacchus* loves the Sunny Hills:
The Yew best prospers in the North, and cold,
The conquered Worlds remotest Swains behold:
See the Eastern *Arabs,* the *Geloni,* these
Countries are all distinguisht by their Trees:

The

The blackest Ebony from India comes,
And from Sabæa Aromatick Gums, &c.

Saffron, Tulips, Anemones, and many other Plants which be propagated by bulbous or tuberous off-sets, require for their melioration, to be planted in a light Soyl, that receives some mixture of fatty earth with it: some commend Cow-dung rotted, above all other soil, to be mixt with other sandy earth for these Plants.

Boggy Plants require, even when they be planted into Gardens, either a natural or artificial Bog, or to be placed near some water, by which there is great improvement to all sorts of Flags, and particularly, as I have observ'd to *Calamus Aromaticus*.

The artificial Bog is made by digging a hole in any stiffe Clay, and filling it with Earth taken from a Bog; or in want of such clay ground, there may be stiffe Clay likewise brought in, and laid to line the hole or pit in the bottom or floor, and the sides likewise, so thick, that the moisture may not be able to get through: Of this sort, in our Physick Garden here in *Oxford*, we have one artificially made by Mr. *Bobart*, for the preservation of Boggy Plants, where being sometimes watered, they thrive as well as in their natural places.

However 'tis true, that there is variety of usuage for Plants of different nature, yet for the generality of Plants, they are best improved by a fat, rich, deep, moist, and feeding Soil; and it is highly his interest that intends a flourishing Orchard, or Kitchin-garden, to improve his ground to the height; divers Flowers reap benefit by the same advantage; as particularly, Carnations and Auricula's; though for these, and some other Plants, the rotten Earth that is usually found in the Bodies of hollow Willow-Trees, is

thought

thought to be a soyl more specifically proper, especially when mixt with other rich Soyl throughly rotten.

That wilde Plants may be meliorated by transplantation into better Soyl, and by being set at greater distances, is no more then what was before noted, and agrees with that of *Virgil, Georg.* 2.

Sponte suâ quæ se tollunt in luminis auras
Infœcunda quidem, sed læta & fortia surgunt
Quippe solo natura subest; tamen hæc quoq; si quis
Inserat, aut scrobibus mandet mutata subactis
Exuerint, Sylvestrem animum, cultuque frequenti,
In quascunq; voces artes, haud tarda sequentur.
Nec non & sterilis quæ stirpibus exit ab imis
Hoc faciet vacuos si sit digesta per agros.
Nunc alta frondes & Rami Matris opacant:
Crescentiq; adimunt fœtus, uruntque ferentem.

Plants that advance themselves t'etherial Air
Unfruitful be, but strong they prove, and fair;
Because they draw their nature from the Soyl:
But these, if any, graft; or shall with toil
Transplant, and then in cultur'd Furrows set
Their wilder disposition they forget:
By frequent culture, they not slowly will
Answer thy labour, and obey thy skill.
So they that spring from Roots, like profit yeild,
If you transplant them to the open Field, (shade,
Which now the Boughs of th' Mother-plant do
And th' Off-sets stop her growth, and make her fade.

The Seed of wilde Cichory that grows every where in the Fields, being sow'd in rich Garden-soyl, is so improved, that we esteem it ordinarily another Plant,

and

and give it the name of Garden-Cichory, though indeed they are the same. But besides the goodness of the ground, and greatness of the distances, there may be some advantage to Field-Plants by changing the Seed, by which action the fermentation is supposed to be augmented in the Ground: Now these changes are either from one kinde to another, as from Wheat and Barly, to Beans and Pease, which is the usual Husbandry of common Fields, or in the same Seed: Of the former way, *Virgil* gives this Precept.

——ibi flava seres mutato sidere farra,
Unde prius latum siliqua quassante legumen,
Aut tenues fœtus Viciæ, tristisque Lupini
Sustuleris fragiles calamos, sylvamque sonantem.
Georg. 1. By Mr. *Ogilby* thus rendred.
—— There changing Seasons thou shalt Barly sow
Where pleasant Pulse with dangling Cods did grow,
Where brittle stalks of bitter Lupines stood,
Or slender Vetches in a murmuring Wood.

Of changing the Seed of the same kinde, besides Field Corn, which is generally changed every third Season at the farthest, examples may be had in Carnations and Gilly-flowers, the Seed of which, being taken from the best Flowers, are much meliorated by alternation and change of Ground; and it is like this Experiment may hold in the seeds of other Flowers.

Another Experiment, is the exossation of Fruit, or causing it to grow without stones or core, for which effect, the grafting of the upper end of the Cyon downwards, hath been asserted to be a certain way: That the Cyon so grafted will grow, I have experience; but whether in time they will produce the
fore-

forementioned effect, I greatly doubt: And if they should, I much miftruft their expectations would not be anfwered, that intended melioration thereby: For the Fruit, certainly by the lofs of the natural Seed, would be very much difpirited, and loofe the generofity and noblenefs of its nature, as Animals do, and as Vegetables fometimes; as particularly I have obferved in Barberries, for I have feen a Tree that bare every year on moft Bunches two forts of Barberries, the one full, and of a deep red; the other of a pale colour, and thin fubftance, and inquiring into the caufe, I found the former to have Stones in them, and the latter deftitute, which were, as I fuppofed, thereby emafculated.

N. 9. *The conclufion of the Treatife, with one or two choice obfervations of the wife and good Providence of God, which may be feen in the admirable make of Vegetables, and fitnefs to their ends, which are not generally taken notice of, but are, with many more, overfeen by men bufie in the affairs of the world.*

It was the fin of the Heathen that they did not rife in their mindes from the contemplation of the beauty of the creatures, to confider how fuch lineaments could be made, and to glorifie thereby the wifdome of the Maker. The particulars are infinite, that ordinarily to a man exercifed in things and thoughts, fuggeft themfelves to avouch Providence, and confute the vanity of the old Epicureans in the fimpleft of their Tenets concerning the framing of this world, of things by a cafuall concurrence of fmall motes intricated

tricated in their motion, by meer chance into such beautifull bodies.

It is no unusuall Theme to treat of the admirable handsomenesse and beauty in the compoiure of divers Vegetables, and to shew how Nature doth γεωμετρεῖν in them, and characterize out such variety of elegant figures, that every plant shall seem to have more of Mathematicall art, than the Knot wherein it is set: And tis generally noted, that Gods Providence is exceeding good in appointing Nature, and making it her end to continue some individualls of every Species for the preservation of the kinde. That likewise the same Providence has approved to its selfe a most excellent wisdom in the choise of most certain meanes, for the attainment of this end, it has been mine, and may be an easie consideration to any other.

For what other end, thought I, are there so many coates, and such cotton vestment to seeds, but to defend their tendernesse? Why such hard stones to other, but to hinder their premature springing, whereby the coldnesse of Winter would kill (as in Aprecots, Peaches, Nectarines, &c.) their tender seedlings? Why is the ground in Woods covered with Mosse, but that Nature intended it as a preservation to seeds fallen upon the Turfe in the violence of Winter Frosts? Why has Nature beset shrubs with prickles, but to defend the tender buds in which the hope of future growth is reposed from the browsing of cattell in the Winter? and that this was the end of Providence in it may be conjectured from hence, because those shrubs which are not all over thorny, have a guard of Thornes directly upon the bud & not else where, as if singularly intended for its security. So tis seen in the Gooseberry, Haw-

Hawthorne, Barbery, Locust, all Roses wild and cultivated that are not all over thorny, so that the thorns are not uselesse excrescencies as some have supposed, but as profitable as boughs or leaves.

Why have those plants that bear no seed with us, as Poplar and Willow, in every bough of any bignesse, a propensity of sending forth Roots, by the occasion of which, each branch is made an entire tree or plant? or if that faculty be wanting, why then is there so great disposition and forwardnesse to propagate themselves by off-sets, as in the Elm, Poplars &c, And where there can be no off-sets, as in Mushroomes, wherefore else has Nature made the plants propagable by the smallest of their shreds and inconsiderable parts? Why else is the Indian Fig, that hath no stalk, propagable by its leaf alone?

Why have plants such an eagernesse to flower and seed, and such an impatience of being disapointed? if you pull off the bud of the Rose it will spring againe, and not only the Rose, but most other fruits and flowers have the same desire to produce their seeds, and have given occasion to Artists to make hence Rules of Retardation.

Why do the Seeds stick close to the Pedall by which they are joyned to the stock untill they are mature and fit for propagation, and then fall off in the most fit season for due preparation to future growth?

Why do those plants that usually die every yeare, yet if they are disappointed of running to seed, continue to survive many years, even so long till they are permitted to run up to leave seed behind them? But that they are appointed by the universall Law of Nature, not to desert their order, till they have produced others after their own kind.

Lastly,

Lastly, Why are many Seeds at their first ripening so exactly fledged with wings, but that by the wind, they may be carried to such places as may be fit wombs to receive and feed them, untill they attaine from the being of seeds the measure and stature of perfect plants.

Another Specimen of the Wisdom of the God of Nature, may be seen in the regular situation of Branches, and the orderly eruption of Buds, upon every Vegetable; for, notwithstanding the report of my Lord *Bacon*, *Nat.° Hist. Cent. 6. Observ.* 588. *That Trees and Herbs in the growing forth of their Boughs and Branches, are not figured, and keep no order, but that when they make an eruption, they break forth casually, where they finde best way in their Bark and Rinde*: I finde my self necessitated to refer that to an exceeding Wisdom, which his Lordship refers to chance and casualty: For if I observe aright both Buds and Leaves, and all eruptions, stand so on every Vegetable, as to serve most fitly for most necessary ends.

As to Leaves, the Learned Doctor *Brown* hath made the Quincunx famous, which may with as great aptness be applyed, and, I think, more universally to the scituation of Buds, or Germens.

This Figure had its name from the numeral Letter V. because the points therein, are the same with the points or Angles in the said Letter, and because that as the Letter is capable of infinite multiplications, so is the Figure, and both in not unlike fashions: The number thus, V. **X. XX. XXX.** the Figure thus, ⸫ ⸫ ⸫ ⸫ ⸫ ⸫

Of

Of this Quincunx I shall propose three sorts. 1. The thicker, as in the Figure *a*. The thinner and less full of points, are either obliquely set, as in the Figure *b*, or more strait, as in the Figure *c*.

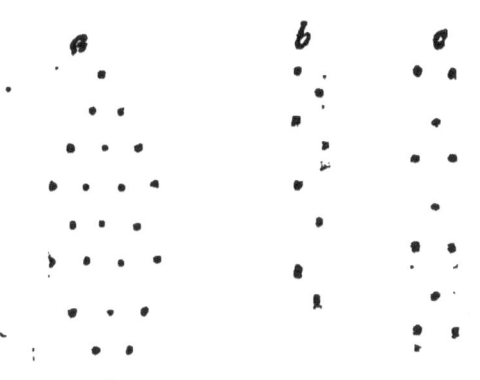

The most thick sort of Quincunx hath its examples rather in Leaves then Buds, for after this manner stand the Leaves upon most Martagons and Lilies, divers Spurges, and Sedums, on which it is most visible, when the Plants run up to Seed. Trickmadam, Spurge-Laurel, Marsh-mallows, when the stock is exceeding ranck and big, for otherwise it is sufficed with the regulations of the third Figure: The leaves of Firre-tree, Pine-tree, &c.

The second, or oblique and single Quincunx, may for the most part be observed, both in the Buds and Leaves that arise from Trees, and such other Plants whose Stalks are round; as in the Oak, Elm, Hasel, Apples, Plums, Cherries, Pears, Willows, Sallows, Osiers, Black-thorn, White-thorn, Coof-berries, Currants, Roses, Fenel, Cichory, Thistles of most sorts, Docks, Bur-docks, Sothern-wood, Rue, Seseli

seli-Æthiopicum, Sweet-maudlin, Common-mercury, Dulcamara.

The third direct and oblong Quincunx is most observed in Plants of a square stalk, as Water-Betony, Fig-wort, Lavander, Mints, St. Johns-Wort, Clowns-All-heal, Rhus-Myrtifolium, Mother-wort, Nep, Colus-Jovis.

Yet tis not unfrequently seen on other Stalks also, as the Sycamore, Elder, Maple, Dog-tree, Ash, Hysope, Nettles, Hemp, Willow-weeds, Tree-Spurge, French-Mercury, Scammony of Montpelier.

And it is to be observed, that in divers of those Plants whose Stalks are set with Joynts, and those Joynts with a beautiful Circle of Leaves, proper to each Plant, contrary to the Quincuncial scituation, the Germens notwithstanding, are found to follow the order of this last mentioned Quincunx, as may be seen in Madder, Goos-grass, Ladies-Bed-straw.

Or if that order be left, yet it is not left to the disadvantage of the Plant, but generally it hath in exchange some other handsome and proper method of Leaves and Buds. Thus *Linaria-Quadrifolia*, hath on each joynt three, four, five or six opposite Leaves, & under each Leaf a Germen, which arise to Branches, uniformly set upon the same round Stalk.

And as to the particular make and frame of those Plants, which in the standing of their Leaves cannot be said to follow the order of any Quincunx, yet they, instead of those elegant Tessellations, are beautified otherwise in their site with as great curiosity. I cannot think of a Plant, according to the ordinary estimation of men, that is more contemptible then that which grows ordinarily in Bogs, or miry Ditches, and is called Great-Horse-tail; yet if any man please to
dis-

disjunctuate the whole, and take particular view both of the parts and conjuncture, they will finde the frame exquisite enough to deserve a better esteem; for both Stalks and Leaves are made up of divers pieces, framed, as it were, in joynt work; all which pieces bear exact proportion each to other; and each receives other by indented terminations, which form very beautiful Coronets on the pieces so received; then at a convenient distance, above each of these Coronets, there ariseth a very beautiful Circle of Leaves, and these very Leaves are made up of hollow pieces articulately, and proportionably joynted, in imitation of the elegancy of the joynts of the Stalk it self.

And generally the Leaves that stand not according to the Quincunx, either stand in joynts, in the fashion of the Burgonion Cross, as on Cross-wort; or in a Circle, as on most sorts of Madder, Ladys-bedstraws, Woodroofs; or in some other profitable, fit and beautiful posture: And though in these creeping and entangled Plants, irregularities are not unfrequently seen, yet even in these irregularities themselves, there often seems to be a greater curiousness and most proper order; as particularly, Madder is generally tetragonal, and notwithstanding its circular border of Leaves, usually sends forth Buds, according to the manner of Mints, and other Plants of a four-square Stalk: This I have sometimes seen in many of its Branches to vary and turn hexagonal, or to have a stalk with six ribs, upon which declension the order of the Germens was thus most fitly altered; upon each rib or angle there was always one leaf, and upon every other rib, a germen under the leaf; which I found so placed, that no one rib did bear the Bud in the two succeeding joynts; so that if in the first joynt, the three Buds stood on the

L   first,

first, the third, and the fifth ribs, then in the second joynt, the Buds stood on the second, the fourth, and the sixth, and so interchangably to the very top.

Now by these scituations of the Buds, according to these Observations, it always is so found necessarily to be, that if two Buds stand on the same joynt, as in the third Quincunx; those that stand on the same heighth, keep always the contrary sides; and further, if the two lowermost stand North and South, the two next immediately above them stand East and West. And in the second, or oblique and single Quincunx, when the Buds stand not two at the same heighth, the second stands on the opposite side to the first, and the fourth to the third; and then likewise, if the first and second stand East and West, the two next above them stand North and South.

I may give notice that to finde these methods, and to expose them to the eye, a profitable way may be to clip off the stalks of the leaves near the Branch, especially in the first and most thick-sort of Quincunx; in the second more single Quincunx, it may not be amiss to slit the Bark and take it off, for it being laid plain and flat, the Quincuncial order will the better appear; the third sort is visible to the eye, as the Plant grows.

Care also must be had, that observation be made on such Plants whose stalks are not twisted, for the twisting of it brings the Leaves and Germens out of order: There may besides these, some other methods appear not here mentioned, but even in them, he that pleases to consider them, I doubt not, will finde constancy for the most part to their rule; or if they have no rule, there may likewise a reason be found why it was good they should be without.

But

But it is most certain, that these are the general methods, and these contrivances of the eruption of Buds, serve for divers excellent ends exceeding fitly, and so are arguments, (how poor and inconsiderable soever these Observations may seem) that they came not out thus by the lucky justlings and stumbling of blinde chance, but by the Providence of a most Powerful, Skilful, and Wise, Artist and Author. For they serve first to procure a fit and proportionable shade for the Stalk and Fruit; neither of which in their tenderness, can endure the scorching Sun-beams; for by keeping this method and order, they communicate their shade to all parts of the Tree or Plant; whereas, should they break out in a disorderly fashion, some parts of the Plant, and some Fruit would be exposed to all weather, where no Buds or Leaves come forth; other parts would be too much shadowed by the two thick eruption of Buds. This order likewise sets out the Boughs and Branches of each Tree into such positions, that one may not easily fret upon another, or gall its neighbor, but grow in a distinct room, every Branch having his proportionable allowance in that circumference which the whole Tree takes up, whereby it may, without any impediment to others, grow to a convenient bigness; otherwise came many Buds out together without method, they could never arrive at any bigness in their future growth, nor attain to good Fruit, or pleasant Leaves and Flowers, but would run out into such thick Crows-Nests, as I have observed sometimes to happen in Plum-Trees by an error or mischance of nature, in the parturition or bringing forth of the Germens. The observation likewise of these methods must needs be of use to the Equilibration and uprightness of

Trees,

Trees, for should all the Boughs break out in one place, or on one side, the heaviness of that side or part, would bend down the body into a crookedness, and deprive it of that uprightness and straitness, which is the most useful site of most Plants; and those that are without these regulations, are generally such as are made to grow upon and twist about other things, and not to bear up themselves, as Bind-weeds, and the like.

And now I am come thus far, there comes into my minde that excellent Animadversion which the most wise King made, when he had considered the several Purposes, Travels, Businesses, Changes and Overtures, which happen to us poor men while we are under Heaven, in their several Seasons; as particularly, in the days of our Birth, and the days of our Death, in the days of our Planting or being Planted, and those of our Plucking, or being Plucked up: When Men get and Increase their Estates, and when they Loose, grow Bankrupt and are undone; in the days of their Jollities, Dancings, Lovings, Wooings and Embracings; as likewise in those cloudy and dull Seasons, when satiety of Enjoyment, indisposition of Body, or other unhappy accidents, has begot Peivishness and Loathing; and when Tears and Mourning contristate all their glory and beauty: Concerning the seasonableness and fitness of all the Estates of men, their conditions, accidents and disasters in their several times, this is his observation, *Eccl.* 3. *That he had seen the travel which God had given the Sons of men to be exercised therewith*, and found, that God by his providence had made every one of the things made, *beautiful in its time: Moreover, that he had set the age in the middle of them, yet so, that no man of them*

can

can finde out the work that God maketh from the beginning to the end.

I shall not Apologize for translating העלם the age or בלם in the middle of them, because I know the words, and methinks the sense and context bear it best, but shall beg leave by a parallelism to apply it to the present matter; the placing, not the timing of things, and to express my thoughts thus: That God has made every thing beautiful in its place, order and scituation, and particularly every part of every Vegetable, and has also set the world so curiously wrought and modell'd, in the middle of us, yet so, that by reason of our various affairs and businesses, and other fancies, no man can finde out the work that God hath made from the beginning to the end.

Lastly, I must beg leave to make the same conclusion and Appendix to the Observation, that the King has there appos'd to his, (*viz.*) That the true and onely use that can be made of those elegancies and beauties which in every aspect suggest themselves unto us, is no other, but that we *Rejoyce in them* and in their Maker, *and do good in this life.* I mean, that we puzzle not our selves over-much, nor discruciate our Spirits to resolve what are the causes, and what the manner of causation of the apparent effects of Gods great power, any further then as our labour may serve for those excellent and firmly together interwoven ends of rejoycing and doing good, and the rather, because of the experiment which this most wise Prince, who was helpt by the great riches of his then puissant Kingdome, (and so not impeded by those wants that usually discomfit private persons in such enquiries) made himself and published concerning his own search, *Eccl.* 1. *That he gave his heart to*

seeke

seek and search out by Wisdome concerning all things that are done under Heaven, and found this to be a sore travell, that God had given the Sons of men to be exercised therewith, And further, That *with much wisdom there is much vexation, and he that increaseth knowledge, increaseth sorrow.*

FINIS.

www.ingramcontent.com/pod-product-compliance
Lightning Source LLC
Chambersburg PA
CBHW022114160426
43197CB00009B/1019